NOTICE AND TAKEDOWN IN EVERYDAY PRACTICE

Jennifer M. Urban, Joe Karaganis
& Brianna L. Schofield

Version 2: Updated March 2017

BerkeleyLaw
UNIVERSITY OF CALIFORNIA
Samuelson Law, Technology
& Public Policy Clinic

THE AMERICAN ASSEMBLY
COLUMBIA UNIVERSITY

Jennifer M. Urban is Clinical Professor of Law at the University of California, Berkeley School of Law and Director of the Samuelson Law, Technology & Public Policy Clinic.

Joe Karaganis is Vice President at The American Assembly at Columbia University.

Brianna L. Schofield worked on this report as a Research Fellow and Clinical Teaching Fellow at the Samuelson Law, Technology & Public Policy Clinic at the University of California, Berkeley School of Law.

THE TAKEDOWN PROJECT

This report is part of the Takedown Project, a global network of researchers studying the role of notice and takedown procedures in addressing conflicts between copyright, privacy, and freedom of expression. More information can be found on the project's website, http://takedownproject.org.

ACKNOWLEDGMENTS AND DISCLOSURES

We are deeply grateful to our survey respondents and interviewees for generously contributing their time and expertise, and for candidly sharing their "on the ground" experiences with notice and takedown.

This work would not have been possible without both data and funding resources for the coding effort. We thank Adam Holland and Wendy Seltzer of Lumen (formerly Chilling Effects) for facilitating access to the Lumen data, which forms the basis for our quantitative work. We are grateful for funding support from Google Inc. as a gift to The American Assembly and from the Sloan Foundation for its support through the Berkeley Law Digital Library Copyright Project. Neither funder directed our approach in any way, and neither funder reviewed any methods, data, results, or reporting before public release. We are also grateful for Nash Information Services' in-kind donation of the OpusData database and expert database customization advice.

We are indebted to the individuals who have generously lent their time, skill, and expertise to the project: Kristoff Grospe for his assistance with the development and management of the qualitative and quantitative studies; Martyn Joyce for building and maintaining our customized database, coding and querying interfaces, and randomization and search algorithms; Bruce Nash of Nash Information Services for help with database and algorithm design; and Nora Broege of UC Berkeley's D-Lab for data preparation and statistical analysis. We are very grateful for the detailed and helpful comments on study design, findings, and drafts offered by Annemarie Bridy, Niva Elkin-Koren, Gwen Hinze, Martin Husovec, Deirdre Mulligan, Bill Rosenblatt, and participants in multiple Takedown Project workshops, the 3rd Global Congress on Intellectual Property Rights and the Public Interest, the 4th Global Congress on Intellectual Property Rights and the Public Interest, the 18th Annual Berkeley Center for Law and Technology and Berkeley Technology Law Journal Symposium "The Next Great Copyright Act", the Department of Commerce Multistakeholder Forum on Improving the Operation of the DMCA Notice and Takedown System, the Chicago-Kent College of Law's Conference on Empirical Research on Copyright Issues, and the Copyright Society of the USA's Copyright and Technology Conference. Mistakes are ours alone.

We thank the exceptional team of graduate students from the University of California, Berkeley School of Law who manually reviewed takedown requests: Sérgio Alves, Jr.; Christina Farmer; Shaina Hyder; Alicia Intriago; Tigist Kassahun; Shweta Kumar; Leighanna Mixter; Smita Rajmohan; and Kirsty Watkins. Lydia Anderson-Dana provided excellent research assistance in the final stages of the project.

IV. STUDIES 2 AND 3: QUANTITATIVE ANALYSIS OF TAKEDOWN REQUESTS 77

EXECUTIVE SUMMARY

BACKGROUND

The Digital Millennium Copyright Act ("DMCA"), passed by Congress in 1998, enshrined a compromise between copyright holders and online service providers ("OSPs") on issues of copyright infringement. Its core feature was section 512, which established a safe harbor mechanism enabling copyright holders to send brief "takedown" requests to OSPs that were to be expeditiously honored, and allowing the targets of these notices to contest requests using a "counter-notice" procedure. Since then, the law and procedure has guided copyright protection on the Internet and has been substantially adopted by several other countries. Yet there has been little empirical research into how the framework actually operates, as applied by rightsholders, OSPs, and targets. The parties have been reluctant to release the private communications (notices and counter notices) and private actions by OSPs that would facilitate empirical research. Further, practices have been affected by the enormous changes in the use of the Internet over the past decade and by the increasing use by large rightsholders of automated "bots" to search and detect possible infringements.

RESEARCH SUMMARY

To shed light on how the system operates in practice eighteen years after the law's passage, this report presents three empirical studies:

- **Study 1** qualitatively documents the ways in which the notice and takedown process has been perceived and operationalized by major U.S. OSPs and rightsholders, based on confidential surveys and in-depth interviews with nearly three dozen OSPs and notice senders.

- **Study 2** quantitatively examines a random sample of takedown notices, taken from a set of over 108 million requests submitted to the Lumen archive over a six-month period (most of which relate to Google Web Search). The quantitative analysis is based on manual review and coding of these notices by the Takedown Project lead researchers and a team of graduate legal researchers at the University of California, Berkeley.[1]

- **Study 3** provides a further detailed quantitative examination of a random sample of notices that were sent to Google in relation to its Google Image Search service, isolated from the same six-month set of takedown requests taken from the Lumen archive, and based on the same manual review and coding process.

Study 1 reveals that OSPs notice and takedown practices and experiences have diverged over the past decade. OSPs split into three broad groups:

[1] Hosted at https://lumendatabase.org/.

1. "DMCA Classic" OSPs—the majority of respondents—for which the volume of notices has remained relatively infrequent and substantial human review of notices is still the norm;

2. "DMCA Auto" OSPs, which receive very large numbers of notices generated by automated systems, have experienced a steep increase in volume in each of the past several years, and have shifted to more automated notice-processing practices;

3. "DMCA Plus" OSPs, which have adopted procedures that go beyond measures required by section 512, including filtering systems, direct takedown procedures for trusted rightsholders, hash-matching based "staydown" systems, and contractual agreements with certain rightsholders that set forth additional protections and obligations for both parties.

There is considerable overlap between DMCA Auto OSPs, which are simply implementing statutory notice and takedown requirements on a massive scale, and DMCA Plus OSPs, which have added additional measures. This research could not consider enough OSPs to discern a clear trend, but it suggests that DMCA Auto tends to "collapse" into DMCA Plus as OSPs attempt to manage very large numbers of takedown requests.

Study 1 reports that, notwithstanding the shifts in the nature of online copyright infringement and responses to it, the law's notice and takedown provisions remain foundational to all the parties interviewed. OSPs consider its safe harbor provisions fundamentally important to their freedom to operate. Rightsholders considered notice and takedown central to their enforcement efforts, though they also expressed frustration with its capacity for dealing with large-scale infringement. Interviews with DMCA Classic OSPs revealed concerns that floods of notices might imminently arrive, forcing them to adopt costly automated systems, though some respondents considered their current situations stable. They also expressed concerns that changes in the law or norms could standardize DMCA Plus measures that raised concerns about their users' expression rights and that they could not afford to implement.

The research highlighted a shift towards professionalization of large-scale copyright enforcement by rightsholders, such as specialized "content protection" teams in trade associations and media companies, and the increased use of third-party rights enforcement organizations ("REOs"). Our qualitative interviews revealed a positive feedback loop that has developed between, on the one hand, the greater use of automated detection and notice sending systems by some rightsholders and REOs, and, on the other hand, of more automated processing by DMCA Auto and DMCA Plus OSPs, which increases the cost and complexity of notice processing for OSPs. Though rightsholders described a variety of methods they use to limit mistakes, both rightsholders and OSPs also noted that automating enforcement can introduce errors. These errors can be hard to catch, as the vast majority of automated claims are not substantially reviewed by humans. Further, OSPs generally described the counter notice procedure as impractical and infrequently used.

Study 2's quantitative analysis revealed deficiencies in notice and takedown procedures, especially automated requests, as all takedown requests in the sample appeared to be automated. About 30% of takedown requests were potentially problematic. In one in twenty-five cases, targeted content did not match the identified infringed work, suggesting that 4.5 million requests in the entire six-month data set were fundamentally flawed. Another 19% of the requests raised questions about whether they had sufficiently identified the allegedly infringed work or the allegedly infringing material. The analysis further identified significant questions related to the availability of potential fair use defense, complaints grounded on improper (non-copyright) claims, and requests sent to defunct web sites.

Study 3's quantitative analysis of the subset of requests sent in relation to Google Image Search revealed different characteristic issues than the Study 2 notices, which were sent in relation to Google Web Search. Study 3 requests were more likely to be sent by less professionalized claimants, including 53% by one individual, and not by automated detection systems. Seventy percent of the requests raised serious questions about their validity, including a significant number related to "improper" subject matter, fair use concerns, copyright ownership issues, and potentially inaccurate identification of the allegedly infringing material.

ANALYSIS AND CONCLUSIONS

The overall picture painted by our research shows that the original process set out in section 512 has evolved into a highly complex ecosystem, with different challenges faced by large incumbents and smaller players, by those in the center of the "copyright wars" and those on the edges, and by those playing different roles in the technology and copyright sectors.

In some of its most basic features, the notice and takedown system is functioning well and section 512's provisions remain central to managing copyright in the online ecosystem. Though it shows strains, the law continues to provide rightsholders with a copyright enforcement alternative that is cheaper and easier to use than lawsuits. Though use of automated systems has reduced human-reviewed processing, non-automated processing is still the norm for most OSPs and continues to work successfully.

The increased use of automated systems by large rightsholders, as well as DMCA Auto and DMCA Plus OSPs, however, raised questions of accuracy and due process. Though rightsholders and OSPs generally use some accuracy checks today, we identified a clear need for better mechanisms to check the accuracy of algorithms, more consistent human review, and a willingness by both rightsholders and OSPs to develop the capacity to identify and reject inappropriate takedown requests. Further, Study 3 indicated that accuracy problems were not limited to automated notices from large senders, and that OSPs may be subject to large numbers of suspect claims, even from a single individual.

The findings also raise issues related to due process at the OSP level. Due process safeguards for targets have largely failed. Study 1 interviews revealed that the counter notice process is rarely used, and has significant structural limitations. The quantitative studies, which showed no use of the counter notice process, reinforced this concern. Without better accuracy requirements for notices, a reasonable ability to respond before action is taken, and an unbiased adjudicator to decide whether takedown is appropriate, counter notice and putback procedures fail to offer real due process protection to targets.

The research also identified problems for rightsholders in responding to large-scale, offshore infringement. In interviews, rightsholders expressed great frustration with extra-territorial infringement-focused sites that do not comply with notice and takedown requirements and are outside the reach of US jurisdiction.

Analyzing the effectiveness of DMCA-mandated procedures in responding to infringing materials on specific sites, balancing copyrights and speech rights, or addressing other concerns is severely limited by the law's lack of requirements for publicly disclosing information on notices sent and OSP responses. This lack of transparency has significant repercussions for OSPs, Internet users, rightsholders, and policymakers. Most respondents lacked awareness of other actors' practices and reasoning. For example, OSPs had no way to gauge whether they were vulnerable to floods of notices requiring expensive, automated responses.

Some secrecy is defensible. Rightsholders expressed concern that publicizing details of their enforcement practices might enable pirates to circumvent them. OSPs worried that revealing their practices could subject them to negative attention from rightsholders, targets, or other OSPs, forcing them to change their practices. Ironically, perhaps, more information and transparency could potentially alleviate this concern by helping the DMCA Classic OSPs better detect problematic notices and the DMCA Auto and Plus OSPs improve their algorithms. Furthermore, secret, algorithmic decision making is difficult for Internet users to penetrate and challenge, rendering their expression rights vulnerable.

DMCA Classic OSPs also expressed the concern that a shift in section 512 compliance standards, or pressure to adopt the practices of large DMCA Plus OSPs, would significantly increase the costs required to receive realistic safe harbor protection. Such a shift could give an advantage to large, well-resourced OSPs and present a barrier to market entry, limiting the robust competition that has enabled the vibrant growth of online services.

RECOMMENDATIONS

Based on our results, we caution policymakers to take into account the complexity of the notice and takedown ecosystem as it is exists today. Neither OSPs nor senders are monolithic groups. What they need from notice and takedown, and the challenges they face, differ widely depending on how they are situated. Changes to the safe harbor or the notice and takedown process that benefit some groups could have highly negative unintended consequences for others. Specifically, policymakers should avoid requirements or effects that would have a disproportionate detrimental impact on OSPs outside the zones of heightened copyright conflict, such as the creation of rules or norms that create barriers to market entry or reduce competition in the OSP sector. Policymakers should also carefully consider the varying needs of different rightsholders, including those with fewer resources or less-sophisticated copyright knowledge. Last but not least, targets' interests should be carefully taken into account, especially as they are unlikely to be present in policy discussions.

Our recommendations for statutory reform focus on making it more difficult for senders to issue questionable notices without risk, and strengthening the ability of targets to respond. They require senders to declare under penalty of perjury that their substantive claims in a takedown notice are accurate, remove the mandatory ten-day waiting period before material goes back up, lower the standard for targets to recover damages from senders who make bogus claims, and raise the penalty for doing so. The recommendations also support reforming the current statutory damages regime to reduce OSPs' fears of outsized liability and current bias toward takedown. The recommendations further suggest requiring notice and counter notice senders to submit notices to a centralized repository, where they can be searched and analyzed.

We strongly recommend avoiding statutory changes that would expand automated practices without much better control against mistake and abuse, or raise the cost of compliance for the vast majority of DMCA Classic OSPs. DMCA Plus measures should remain entirely voluntary.

Beyond statutory reform, knowledge-sharing and best practices can fill gaps and improve operation. Our qualitative interviews identified a number of such practices. Accordingly, our recommendations for rightsholders emphasize both human and machine methods to help limit mistakes and misuse. We encourage rightsholders to work with OSPs to streamline processing and limit overbroad removal.

Similarly, OSPs that use automated notice-processing systems should develop mechanisms to flag questionable notices for human review and reduce overbroad takedowns. Good practices include developing better filters that identify both flawed notices and questionable senders, and routine spot checks. OSPs should provide senders with educational materials and guidance about appropriate takedown requests, and provide targets with educational materials and an easy-to-use counter notice function.

Stakeholders and government agencies should also develop informational resources and guidelines for senders and targets on copyright law, the scope and requirements of the notice and takedown regime, and how to send notices and counter notices. The materials could be hosted by a neutral government entity and accessible to all notice and counter notice senders.

I. INTRODUCTION AND SUMMARY OF FINDINGS

In 1998, Congress passed a then-obscure law with enormous implications for copyright holders, Internet speakers, and online service providers. It defined remedies available to rightsholders and responsibilities of online service providers ("OSPs")—and indeed shaped the future of the Internet itself. In the eighteen years since it was passed, the "notice and takedown" system established by section 512 of the Digital Millennium Copyright Act[2] ("DMCA") has become a primary tool for raising and resolving copyright disputes in the United States, and has served as a model for other countries.

The fundamental issue originally settled by section 512 was whether the new types of Internet intermediaries—OSPs—that hosted or transmitted material from users "without modification" would be treated as publishers of that material, and therefore liable for copyright infringement.[3] Copyright holders argued that traditional publisher liability provided the right model for the new intermediaries to address the vastly expanded capacity for copyright infringement on the Internet. OSPs, in turn, argued that many of the new services enabled by the Internet precluded the type of editorial involvement on which publisher liability has relied. What was possible for a magazine publisher or newsroom editor, they contended, was not scalable to thousands or millions of user-generated posts, comments, or data transfers. They argued that the Internet's emerging radical expansion of expression—since anyone with a connection[4] and an account with an OSP could now reach a worldwide audience without relying on a traditional publisher—could be crushed under OSPs' fears of liability for their users' infringement.

Section 512 embodied a compromise giving OSPs a "safe harbor" from secondary liability for their users' copyright infringement. In return the OSPs were required to implement certain features to protect copyright holders, most notably so-called "notice and takedown" procedures. Under these procedures, copyright owners can get infringing materials removed from online sites by sending brief "takedown notices" to OSPs, without the expense and hassle of filing a lawsuit. Protections for the expression interests of users were added at the last minute by Senator John Ashcroft, in the form of a "counter notice" process that gave the target of a notice the ability to respond and request "putback" by the OSP. In this way, the DMCA's drafters attempted to strike a balance among the remedies available to rights holders, the responsibilities of OSPs, and the protections afforded to targeted users.

---

2 The On-Line Copyright Infringement Liability Limitation Act (OCILLA)—commonly known as Section 512 of the DMCA—is codified as Title II of the DMCA at 17 U.S.C. § 512 (2012). Title I of the DMCA—which covers anticircumvention—is unrelated.

3 For a comprehensive account of the passage of the DMCA, *see* Jessica Litman, Digital Copyright: Protecting Intellectual Property on the Internet (2000).

4 Despite the importance of this shift, however, it is far from complete. The digital divide remains significant. By the end of 2015, only 7% of households in the least developed countries and 34% of households in developing countries will have Internet access, compared to 80% in developed countries. *See* Int'l Telecomm. Union, ICT Facts & Figures (2015), http://www.itu.int/en/ITU-D/Statistics/Documents/facts/ICTFactsFigures2015.pdf.

But today, the compromise established by section 512 may be unraveling. The law is at the center of hot disputes between the movie, music, and publishing industries on the one hand, and intermediary OSPs, on the other. These disputes largely arise from forces the legislators could not have foreseen in 1998. Since the law's passage, the Internet has grown massively and undergone tremendous technological and social change. Google was launched in 1998, the year section 512 was passed; Napster followed the next year. Facebook and other social networks followed several years later. These social networks, along with YouTube and other platforms, have emerged as dominant content purveyors and major speech platforms that have radically democratized expression. The nature of section 512's "notice and takedown" procedures has also changed. Faced with large-scale infringement, large corporations now use automated "bots" to search for copyright violations and generate millions of automated "takedown" notices to OSPs. While this allows some copyright owners to police their copyrights on today's Internet, relying on machines to make decisions about sometimes-nuanced copyright law raises questions about the effect on expression.

In years since they took effect, and especially since the arrival of bots, the notice and takedown provisions have been used by rightsholders countless—but likely billions—of times. Still, OSPs consider section 512's liability protections so important that, in the words of one our respondents, the law stands as the "foundational legal enabler of online services." This influence extends well beyond US-based copyright disputes. Section 512's basic procedural framework has been exported into international agreements and laws, and treated as a template for OSP policies covering topics other than copyright, from trademark infringement to abusive online speech.

But does it work? Eighteen years after enactment of the DMCA, the question is still hotly debated. Despite the enormous changes since the law was passed, there have been few empirical studies of how notice and takedown actually works in practice—largely because the relevant data has been so hidden from public view and so politically sensitive to the parties involved. Reviews by Urban and Quilter (2006), Quilter and Heins (2007), and the recent statistical inquiry into notices by Seng (2014)[5] largely exhaust the empirical research literature on the topic.[6] The most recent and detailed source is a US Department of Commerce task force "green paper" documenting stakeholders' experiences addressing online copyright issues, including stakeholders' interaction with notice and takedown under the DMCA.[7]

The record for section 512 adds little to the available data. Because it relies on a series of private notices and actions by private parties, notice and takedown largely operates without public visibility into the practices of rightsholders, OSPs, and alleged infringers. So while

---

5 *See* Jennifer M. Urban & Laura Quilter, *Efficient Process or "Chilling Effects"? Takedown Notices Under Section 512 of the Digital Millennium Copyright Act,* 22 Santa Clara Comp. High Tech. LJ 621 (2006); Laura Quilter & Marjorie Heins, Brennan Ctr. for Justice, Intellectual Property and Free Speech in the Online World: How Educational Institutions and Other Online Service Providers Are Coping with Cease and Desist Letters and Takedown Notices (2007); Daniel Seng, *The State of the Discordant Union: An Empirical Analysis of DMCA Takedown Notices*, 18 Va. J.L. & Tech 369 (2014).

6 There are also a few recent, but generally narrower empirical reviews of copyright takedown activity, *See, e.g.*, Bruce Boyden, The Failure of the DMCA Notice and Takedown System: A Twentieth Century Solution to a Twenty-First Century Problem (2013), http://cpip.gmu.edu/wp-content/uploads/2013/08/Bruce-Boyden-The-Failure-of-the-DMCA-Notice-and-Takedown-System1.pdf; Kris Erickson et al., Copyright and the Economic Effects of Parody: An Empirical Study of Music Videos on the YouTube Platform and an Assessment of the Regulatory Options (2013), https://www.gov.uk/government/uploads/system/uploads/attachment_data/file/309903/ipresearch-parody-report3-150313.pdf.

7 Dep't of Commerce Internet Policy Taskforce, Copyright Policy, Creativity, and Innovation in the Digital Economy (2013).

there is controversy about notice and takedown, including a more-or-less continuous supply of court decisions and legal commentary, there is relatively little information about how the system operates in practice and how it affects stakeholders. The technology policy press documents some of the more striking instances of mistaken takedowns, but the anecdotal quality of much of this reporting makes it difficult to determine how prevalent such failures are and whether they reflect systemic issues.[8]

This report helps fill the gap in empirical studies on the effects of notice and takedown regulations. Our qualitative and quantitative research yielded surprising results. And it suggests new ways of conceiving and putting into practice policies that build upon section 512 to more economically and effectively resolve future conflicts.

The primary obstacle to more systematic accounts of the notice and takedown process has been the scarcity of detailed reporting by the major stakeholders—including all of the rightsholder groups[9] and third-party rights enforcement organizations ("REOs")[10] and all but a handful of OSPs. Existing analyses tend to focus on notices sent to Google because Google is the only service that has publicly reported on notice and takedown over a substantial period of time. The record began in 2002 with the archiving of notices to Google Web Search with the Chilling Effects (now Lumen) project.[11] In 2011, Google began including online reporting of aggregate statistics for some of its services in a regular "Transparency Report."[12]

To the existing literature, this report adds three empirical studies that document the notice and takedown process as it works today. Using both qualitative and quantitative methods, the research provides a detailed account of notice and takedown, first from the perspective of stakeholders who interact with the system, and then through independent reviews of takedown requests.

- **Study 1** qualitatively documents the ways in which the notice and takedown process has been operationalized by major U.S. OSPs and rightsholders, based on confidential surveys and in-depth interviews with nearly three dozen OSPs and notice senders.

---

8 *See, e.g.*, Nate Anderson, Ars Technica, http://arstechnica.com/author/nate-anderson/ (last visited Feb. 5, 2016); Michael Masnick, Techdirt, https://www.techdirt.com/user/mmasnick (last visited Feb. 5, 2016); Ernesto Van der Sar, TorrentFreak, http://torrentfreak.com/author/ernesto/ (last visited Feb. 5, 2016).

9 One researcher, however, reported on numbers provided by the MPAA. *See* Boyden, *supra* note 6, at 3.

10 REOs are third-party services rightsholders hire to investigate infringement and send takedown notices.

11 *See* David F. Gallagher, *New Economy; A Copyright Dispute with the Church of Scientology is Forcing Google to Do Some Creative Linking*, N.Y. Times (Apr. 22, 2002), http://www.nytimes.com/2002/04/22/business/new-economy-copyright-dispute-with-church-scientology-forcing-google-some.html?src=pm.

12 *Transparency Report*, Google, http://www.google.com/transparencyreport/removals/copyright/?hl=en (last updated Nov. 5, 2015). Further, even Google only reports on some of the takedown activity targeting its services. Notably, it does not report on YouTube. *See id.* Google's online takedown reporting has more recently been adopted by a growing number of OSPs, though still only a handful compared to the wider landscape. If they grow over time, these "transparency reports" may help researchers develop a broader picture. *See, e.g.*, Mega, Mega Transparency Report (2015), https://mega.nz/MegaTransparencyReportMarch2015.pdf; Reddit, Reddit Transparency Report: Requests for User Information and for Removal of Content (2015), https://www.redditstatic.com/transparency/2014.pdf; *Microsoft Transparency Hub*, Microsoft, https://www.microsoft.com/about/corporatecitizenship/en-us/transparencyhub/crrr/ (last visited Feb. 5, 2016); *Copyright Notices*, Twitter https://transparency.twitter.com/copyright-notices (last visited Feb. 5, 2016); *Requests for Content Alteration & Takedown*, Wikimedia, https://transparency.wikimedia.org/content.html (last visited Feb. 5, 2016); *Intellectual Property*, WordPress, https://transparency.automattic.com/intellectual-property/ (last visited Feb. 5, 2016).

It also examines how stakeholders characterize their experience of the process as currently framed under US law.

- **Study 2** quantitatively examines a random sample of takedown notices taken from a set of over 108 million requests submitted to the Lumen[13] archive over a six-month period (most of which relate to Google Web Search). The quantitative analysis is based on manual review and coding of these notices by the Takedown Project lead researchers and a team of graduate legal researchers at the University of California, Berkeley.

- **Study 3** provides a further detailed quantitative examination of a sample of notices that were sent to Google in relation to its Google Image Service, isolated from the same random sample of takedown requests taken from the Lumen archive and based on the same manual review and coding process.

Study 1 showed that the DMCA is deeply embedded in the practice and policies of both OSPs and rightsholders, and that its liability protections remain central to OSPs' sense of their freedom to operate. OSPs described the process as fundamental to their survival in an environment in which high statutory penalties for copyright infringement could easily sink companies. Rightsholders agreed that notice and takedown is fundamental to their enforcement strategies, but uniformly described its provisions as inadequate for addressing large-scale online infringement. All respondents agreed that the provisions are, as one put it, "woven into the fabric" of the policies, practices, and physical infrastructure that OSPs and rightsholders have built to manage online infringement.

Beyond this general agreement, the practices of both groups showed more variation and nuance than is generally reported. The study revealed strikingly divergent practices by different actors in the notice and takedown ecosystem. As noted, in response to internet-scale infringement concerns, some rightsholders have transformed their notice sending practices by adopting automated systems to detect infringing content and send takedown notices. For some OSPs, this automation increased the annual number of notices they received to hundreds of thousands or even millions of requests. Some OSPs responded by sacrificing human review of the vast majority of takedown requests and deploying their own automated processing methods to accomplish takedown more efficiently. In the study, OSPs using these **"DMCA Auto"** practices described feeling a great deal of pressure to maintain compliance with takedown demands at ever-increasing costs. In some cases, they have moved beyond reactive automation to proactive methods—filtering, for example—that go beyond the DMCA framework of requirements. These are entitled **"DMCA Plus"** practices.[14]

This shift to automation by some rightsholders and OSPs tends to receive attention in policy debates over section 512. However, Study 1 also showed that for many OSPs, notice and takedown continues to operate largely as it has since the DMCA took effect, *without* large-scale notice sending and handling. These **"DMCA Classic"** OSPs—often, but not always, companies outside of the contested music, video, and search areas—receive relatively small numbers of notices and subject them to review by human teams. Some DMCA Classic OSPs

[13] Hosted at https://lumendatabase.org/.

[14] A few OSPs identified a third group, which is even further afield from section 512's original framework. These **"Para DMCA OSPs"** likely would not incur secondary liability for user copyright infringement under U.S. law—they may include advertising and payments providers, for example. Though we spoke directly only with OSPs that fall under the existing section 512 framework, enforcement efforts increasingly extend to these Para DMCA OSPs, and they often make up part of an enforcement ecosystem within which DMCA-covered OSPs are acting.

viewed this state of affairs as stable. Others considered their situations more vulnerable, and worried that future deluges of notices could force them to abandon human review and adopt DMCA Auto or Plus measures. Notably, DMCA Classic OSPs considered the costs of DMCA Auto or DMCA Plus practices high enough to limit market entry and success, and feared legal or norm-based pressure to implement these measures.

Taken as a whole, Study 1 suggests some avenues for potential reforms and practice updates, but with a strong caveat: interventions risk creating unexpected consequences if they do not take into account both the structural investments OSPs and rightsholders have made in response to section 512 and the different practices of small and large rightsholders, small and large OSPs, and DMCA Classic OSPs as well as DMCA Auto and Plus OSPs.

Studies 2 and 3 take a quantitative approach. In these studies, we analyze a representative sample of takedown requests sent to Lumen over a six-month period—more than 108 million requests in total. Using detailed hand coding of two randomized samples from the dataset, we explore the integrity of the notice and takedown process in this system. Study 2 is based on a randomized sample of the entire six-month cohort of notices, the vast majority of which were sent to Google. It delves directly into automated noticing. Automated notices to Google's Web Search service are such an overwhelming feature of the overall cohort that further sampling was required to isolate other services that may exhibit different features. Accordingly, Study 3 focuses on a separate randomized sample of only the notices sent to Google's Image Search service.[15]

Again, we observed a divergence, in this case between the Study 2 notices, all of which were sent Google Web Search, and the notices sent to Google Image Search observed in Study 3. Takedown requests to Google Web Search number in the millions per week, are overwhelmingly sent by or on behalf of large entertainment companies with valuable properties, and largely target file-sharing sites and other sites where large-scale infringement takes place. Takedown notices to Image Search, however, are several orders of magnitude fewer, appear to be sent mainly by individuals and small businesses, and target a wide range of expression posted on social media sites, personal websites, blogs, and forums. Defects in the takedown process were unfortunately common in both studies, but the issues differed.

Study 2 uncovered issues arising from automated decision-making and the vast scale on which the requests are sent and processed.

- 31% of takedown requests were potentially problematic, including both fundamentally flawed requests and "questionable" requests:
    - One in twenty-five of the takedown requests (4.2%) were fundamentally flawed because they targeted content that clearly did not match the identified infringed work. This translates to approximately 4.5 million requests[16] across the entire six-month set that could be expected to suffer from this problem.

---

[15] We have not been able to analyze all potential subsets of the data, but we have made the data and coding and querying engines available to other researchers. Some are currently preparing additional analyses. For example, Niva Elkin-Koren, Sharon Bar-Ziv, and Nati Perl at the Haifa Center for Law & Technology are currently studying all the notices in the cohort sent to .il (Israel) domains.

[16] The margin of error for our sample is +/-2.29 with 95% confidence, so we can expect a range from 2 million to 7 million in the entire 108.3 million.

- Nearly a third (28.4%) had other characteristics that raised questions about their validity. (Where a single request presented multiple potential issues we counted it as "questionable" only once.)
  - Greater than 19% of takedown requests raised questions about whether they complied with the most substantive statutory requirements: sufficiently identifying the allegedly infringed work or the material alleged to infringe. These are fundamental requirements, as identifying the works in question is necessary to evaluate claims and to take down material.
  - About one in fifteen (6.6%) of takedown requests raised questions of potential fair use defenses.
  - Smaller numbers of takedown requests (2.3%) complained of subject matter inappropriate for DMCA takedown, such as trademark or defamation claims.
  - A few parties continued to send takedown requests targeting links that led to sites that were long defunct, demonstrating a lack of quality control in automated methods.

Where the Study 2 takedown requests appeared to reflect the challenges of automation, the Study 3 Image Search requests presented quite different characteristic issues. These requests tended to be sent by less professionalized claimants—individuals and small businesses—and did not appear to be automated. Study 3 requests raised even more substantive questions about the underlying claim, perhaps because their senders were more likely to misunderstand or misuse the process than the professional Study 2 senders.

- Strikingly, nearly 53% of the Google Image Search takedown requests were from one individual sender, Ella Miller.[17] All of these requests appeared to be improper subject matter for DMCA takedown—none were copyright complaints.
- Overall, including the Miller requests, seven out of ten (70%)[18] of the Google Image Search takedown requests presented serious questions about their validity.
- Even without the Miller requests, 36.3% of the remaining Google Image Search takedown requests were questionable. These broke down into several categories:
  - 15.1% raised questions about the subject matter of the claim (this increases to 60% when the Miller notices are included);
  - 11.5% exhibited characteristics that suggested possible fair use defenses;
  - 6.1% presented ownership issues;
  - 2.9% presented questions about whether the sender had identified the allegedly infringing material; and
  - a small number (.9%) targeted material likely to be in the public domain.

[17] This is a pseudonym.

[18] For purposes of calculating the total number of questionable requests, a request that has multiple questionable characteristics is not counted more than once.

Because Google Image Search receives so many fewer requests than Google Web Search, these problems translate into problems with thousands (rather than millions) of requests in the entire six-month set—still a sobering number. Further, the Image Search requests were much more likely to implicate individual expression by targeting posts on social media, personal websites, and blogs, rather than the file-sharing or cyberlocker sites more commonly targeted in requests sent to Google Web Search.

Together, the three studies provide the broadest empirical analysis of DMCA notice and takedown of which we are aware to date. The studies, in Sections III and IV, are bookended by Section II, which provides background on the legal framework for notice and takedown and by Sections V and VI, which analyze the findings, present recommendations for reform, and conclusions.

II. LEGAL BACKGROUND AND NOTICE AND TAKEDOWN'S EXPANDING REACH

As the studies consider how section 512's liability limitations and notice and takedown regime work in practice today, we provide a brief background on its purpose, structure, application by courts, and the ongoing debates over its allocation of responsibility between copyright owners and OSPs. Other, more detailed accounts of the DMCA's history and its complicated and evolving history of judicial interpretation are available.[19] For purposes of this report, we limit the discussion to those aspects of the statute and court cases interpreting it that appeared most clearly to affect current and evolving practices.

We also include some discussion of section 512's overall influence through the remarkable expansion of its notice and takedown framework into other jurisdictions and other areas of contested speech.

A. STATUTORY LANGUAGE AND DEBATING THE ALLOCATION OF RESPONSIBILITY

Under the DMCA's statutory framework, eligible OSPs receive safe harbor from secondary copyright liability for infringement by their users if they meet several conditions set out in the statute.[20] Intermediaries that provide four types of online services are eligible for protection: (a) transitory digital network communications, where an OSP acts as a "mere conduit" in providing Internet access;[21] (b) system caching;[22] (c) information residing on systems or networks at the direction of a user (including hosting);[23] and (d) providing information location tools, such as links.[24]

To obtain the benefit of the safe harbors, OSPs must not have "red flag knowledge" of infringement on their systems.[25] They must also comply with two threshold conditions that apply to any of the four types of OSPs, and each type of OSP must also comply with a set of specific qualifying conditions.

First, all OSPs must adopt and reasonably implement "a policy that provides for the termination in appropriate circumstances of subscribers and account holders" who are

---

[19] *See, e.g.*, Jessica Litman, Digital Copyright: Protecting Intellectual Property on the Internet (2000); William Patry, Moral Panics and the Copyright Wars (2009). A large number of scholarly papers also trace the history in shorter form; for example, Urban & Quilter, *Efficient Process*, *supra* note 5.

[20] The Digital Millennium Copyright Act, Pub. L. No. 105-304, 112 Stat. 2860 (Oct. 28, 1998). Title II (codified at 17 U.S.C. § 512 (2006)).

[21] 17 U.S.C. § 512(a).

[22] 17 U.S.C. § 512(b).

[23] 17 U.S.C. § 512(c).

[24] 17 U.S.C. § 512(d).

[25] 17 U.S.C. § 512(c)(1)(A)(ii).

"repeat infringers", and inform its subscribers and account holders of this. Second, all OSPs must accommodate and not interfere with "standard technical measures" that rightsholders use to identify or protect copyrighted works.[26]

Hosting platforms and search engines/location tool providers must comply with three additional conditions. First, they must not possess *actual* knowledge that material or an activity using material on the intermediary's system or network is infringing,[27] and must not be aware of facts or circumstances from which infringing activity is apparent ("red flag" knowledge). Upon obtaining such knowledge or awareness, the OSP must expeditiously remove, or disable access to the allegedly infringing material, or it will lose statutory safe harbor protection.[28] Second, OSPs must not have received a financial benefit that is directly attributable to infringing activity where the OSP has "the right and ability to control" that activity.[29] Third, the OSP must implement the "notice and takedown" procedure. Upon receiving a valid notice from a copyright holder or its agent of specific allegedly infringing content that is posted on the hosting platform, or to which location tool providers have linked, the OSP must act expeditiously to remove or disable access to the identified material.[30] To be valid, triggering an obligation for an OSP to respond, takedown notices must contain specified information, including the location of the allegedly infringing material, identified with specificity.[31] Courts have found that notices that do not identify the allegedly infringing information with sufficient specificity do not provide OSPs with knowledge of infringement that would subject them to liability.[32]

The DMCA regime includes several procedural measures designed to provide due process to Internet users whose material is targeted in the takedown notice, and to limit potential mistaken or overbroad content removal. First, targets can submit a counter notice to the OSP that has taken down their content in certain circumstances.[33] On receipt of a valid counter notice, OSPs must replace the allegedly infringing content within ten to fourteen days after its removal if the copyright complainant has not filed a copyright infringement lawsuit against the counter-noticing Internet user. (The practical effect of this "putback" requirement is unclear, however, as enforcing it would require a lawsuit by the target against the OSP. Many OSPs attempt to immunize themselves from such suits in their contractual terms of service with users.) Second, to protect against misuse of the notice and counter notice procedures, the DMCA provides a right of action to recover damages, and costs (including attorneys' fees) for any party's knowing material misrepresentation in a notice

---

26 17 U.S.C. § 512(i)(2).

27 17 U.S.C. § 512(c)(1)(A); § 512(d)(1).

28 17 U.S.C. § 512(c)(1)(A); § 512(d)(1).

29 17 U.S.C. § 512(c)(2); § 512(d)(2).

30 17 U.S.C. § 512(c)(1)(C).

31 17 U.S.C. 512(c)(3). Where an OSP receives a notice that does not comply with the requirements of 17 U.S.C. § 512(c)(3)(A)(i)-(vi), but *substantially* complies with (ii), (iii), and (iv), the notification shall not be deemed to trigger knowledge if the OSP promptly attempts to contact the complainant to assist in the receipt of a notification that substantially complies with all of the provisions (See 17 U.S.C. § 512(c)(3)(B)(ii)).

32 Perfect 10, Inc. v. CCBill, LLC, 488 F.3d 1102, 1114 (9th Cir. 2007); UMG Recordings, Inc. v. Veoh Networks Inc., 665 F. Supp. 2d 1099, 1108-09 (C.D. Cal. 2009), *aff'd sub nom.* UMG Recordings, Inc. v. Shelter Capital Partners LLC, 667 F.3d 1022 (9th Cir. 2011), *opinion withdrawn and superseded on reh'g*, 718 F.3d 1006 (9th Cir. 2013), *and aff'd sub nom.* UMG Recordings, Inc. v. Shelter Capital Partners LLC, 718 F.3d 1006 (9th Cir. 2013); Hendrickson v. eBay, Inc., 165 F. Supp. 2d 1082, 1093 (C.D. Cal. 2001).

33 17 U.S.C. § 512(g)(2)-(3).

or counter notice that results in content being improperly removed or restored.[34] Finally, the regime contains an important general limitation intended to reduce incentives for OSPs to actively police expression on their systems: the safe harbors cannot be conditioned on a requirement that an OSP monitor its service, or affirmatively seek out facts indicating infringing activity.[35]

In developing the notice and takedown process, Congress intended that the safe harbor regime would "provide strong incentives for service providers and copyright owners to cooperate to detect and deal with copyright infringements."[36] To achieve that end, Congress divided the burdens of compliance between OSPs and copyright owners. Congress placed on Internet intermediaries the burden of responding to valid takedown notices by "expeditiously" removing or disabling access to the identified allegedly infringing content. Congress placed the burden on copyright holders to identify infringing material because it considered that they know what material they own, and "are thus better able to efficiently identify infringing copies than service providers [...] who cannot readily ascertain what material is copyrighted and what is not."[37] Courts have since affirmed that the DMCA notice and takedown provisions follow longstanding copyright law by "plac[ing] the burden of policing ongoing copyright infringement—identifying potential infringing material and adequately documenting infringement—squarely on the owners of the copyright."[38]

Legal and policy arguments over the DMCA's effectiveness and fairness have been ongoing since its inception and center on a few key issues that allocate responsibilities and costs between the parties. First, arguments over effectiveness have centered on whether its notice and takedown procedures can scale up to address determined, large-scale piracy. Second, various stakeholders have advanced arguments over fairness. Rightsholders have challenged the requirement that they identify individual infringements and have pushed in court for more expansive readings of when an OSP's level of knowledge about infringement or its activities take it out of the safe harbor. OSPs have, largely successfully, defended themselves in court by arguing that rightsholders must identify specific instances of infringement when sending notices. Finally, arguments on behalf of target interests have centered on whether the statute sufficiently manages possible overbreadth in takedown.

Debates over OSPs' and rightsholders' relative responsibilities have come together in recent court cases and policy debates centering on when "red flag" awareness of infringement occurs and whether section 512 supports the view that OSPs have a "notice and staydown" obligation in order to benefit from the statutory safe harbor. Depending on what counts as "red flag" awareness, OSPs' and rightsholders' responsibilities and costs could vary widely.

---

[34] 17 U.S.C. § 512(f).

[35] 17 U.S.C. § 512(m).

[36] S. REP. No. 105-190 at 20; H.R. REP. No. 105-551, pt. 2, at 49.

[37] S. REP. No. 105-190 at 48; H.R. REP. No. 105-551, pt.2 at 57-58.

[38] Perfect 10 v. CCBill, 488 F.3d 1102, 1113 (9th Cir. 2007).

Accordingly, the standard has been litigated nearly continuously since it was instated.[39] Although in theory "something less than a formal takedown notice may… establish red flag knowledge,"[40] in practice "red flag" knowledge generally arises when OSPs receive notice of specific infringing materials. For example, in *Viacom v. YouTube*, a leading Second Circuit Court of Appeals case, the court held that to lose the safe harbor, an OSP must have knowledge or awareness of *specific, identifiable* instances of infringing activity—not just generalized knowledge of the possibility of infringing activity on the service.[41] The court reasoned that expeditious removal of infringing content is only possible when an OSP knows with particularity which items to remove.[42] The court further held, however, that the "willful blindness" doctrine could be applied to establish red flag knowledge where an OSP makes a "deliberate effort to avoid guilty knowledge."[43]

Even absent the requisite knowledge, where an OSP has the "right and ability to control" infringing activity it must not "receive a financial benefit directly attributable to infringing activity" to be eligible for the safe harbor.[44] The *Viacom* court set quite a high bar for determining that an intermediary has a level of control that disqualifies it from safe harbor protection. Intermediaries must have "something more" than a mere contractual right and ability to terminate user accounts; rather, they must have substantial, demonstrated influence

---

39 *See, e.g.*, *Perfect 10*, 488 F.3d at 1114 (finding that notices that do not substantially comply with 512(c)(3) do not impart red flag knowledge, use of "illegal" or "stolen" in domain name does not create red flag knowledge); Io Grp. v. Veoh Networks, 586 F. Supp. 2d at 1148; Corbis Corp. v. Amazon.com, Inc., 351 F. Supp. 2d 1090, 1108-09 (W.D. Wash. 2004) ("[A]pparent knowledge requires evidence that a service provider "'turned a blind eye to 'red flags' of obvious infringement'" (quoting H.R.Rep. No. 105–551, pt. 2, at 57); (finding that neither general knowledge of infringement on the site nor third party notices are enough to constitute a "red flag"); Hendrickson v. eBay, Inc., 165 F. Supp. 2d 1082, 1093 (C.D. Cal. 2001) (a notice which does not comply with 17 U.S.C. § 512(c)(3)(A)(ii)'s adequate identification requirement is not considered when evaluating whether the OSP has actual or "red flag" knowledge).

40 Capitol Records, Inc. v. MP3tunes, LLC, No. 07 CIV. 9931 WHP, 2013 WL 1987225, at *4 (S.D.N.Y. May 14, 2013).

41 Viacom Int'l, Inc. v. YouTube, Inc., 676 F.3d 19, 30-31 (2d Cir. 2012).

42 *Id.* at 30; *see also* UMG Recordings, Inc. v. Shelter Capital Partners LLC, 718 F.3d 1006, 1021-22 (9th Cir. 2013) (declining to adopt "a broad conception of the knowledge requirement" and holding that the safe harbor requires "specific knowledge of particular infringing activity); Capitol Records, Inc. v. MP3tunes, LLC, 821 F. Supp. 2d 627, 644 (S.D.N.Y. 2011), *on reconsideration in part*, No. 07 CIV. 9931 WHP, 2013 WL 1987225 (S.D.N.Y. May 14, 2013) (holding that general awareness of rampant infringement is not enough to disqualify a service provider of protection).

43 *Viacom*, 676 F.3d at 35 (stating that the willful blindness doctrine may be applied, in appropriate circumstances, to demonstrate knowledge or awareness of specific instances of infringement under the DMCA). The details of how this standard might be applied to YouTube was left unresolved in *Viacom* when the case settled. In *Capitol Records, LLC v. Vimeo, LLC*, the court declined to certify for interlocutory appeal of the Court's holding that the OSP must be willfully blind to specific instances of infringement to lose safe harbor protection. 972 F. Supp. 2d 537, 555 (S.D.N.Y. 2013). The *Vimeo* case, currently on appeal in the Court of Appeals for the Second Circuit, also opens a question as to whether an OSP viewing a user-generated video on its platform that contains "all or virtually all of a recognizable, copyrighted song" could lead to red flag knowledge. *Id.* at 553. The court certified for interlocutory appeal the question of whether, under *Viacom v. YouTube*, a service provider's viewing of a user-generated video containing all or virtually all of a recognizable, copyrighted song may establish "facts or circumstances" giving rise to "red flag" knowledge of infringement. *Id.*

44 17 U.S.C. § 512(c)(1)(B), (d)(2).

over their users, either through exercising a high level of control over them, or by engaging in purposeful conduct that encourages them to infringe.[45]

Applying this standard, the Ninth Circuit Court of Appeals refused to construe the "right and ability to control" as requiring hosting platforms to engage in ongoing monitoring to ensure that removed allegedly infringing content "stayed down." In *UMG v. Veoh*, it declined to accept the argument that in order to retain the statutory safe harbor, the platform should have gone beyond removing content identified in notices, and programmed its hash filtering technology to ensure that any attempt to repost the same content would be blocked.[46] The court found that the defendant, Veoh, did not have an obligation to police its service to this extent, and that the UMG's interpretation of "right and ability to control" as a form of "notice and staydown" was not supportable. According to the court, that interpretation would have run afoul of the section 512(m) prohibition against conditioning safe harbor availability on a monitoring obligation, and would have transferred the burden of identifying allegedly infringing content to from rightsholders to OSPs, contrary to Congress' intent and a prior ruling in the *CCBill* case.[47]

These debates over knowledge, culpability, and sufficient notice play out in practice daily. Some copyright holders have publicly decried striking the balances for knowledge and "right and ability to control" where courts have thus far, arguing that OSPs should share more burden of "detecting and dealing with" online infringement.[48] Failure to persuade courts in several cases has not ended litigation or political efforts on the issue. OSPs, for their part, have argued that section 512 was predicated on rightsholder identification of infringement, and that they cannot reasonably be expected to detect infringement without notice, as only rightsholders are in a position to know the rights and licensing history of their works.

B. EXPANSION OF THE DMCA FRAMEWORK

Whatever the conclusion on whether the DMCA succeeds at balancing copyright interests online, its basic notice and takedown framework has been replicated or adapted many times. Indeed, it has become the go-to model for those attempting to solve any number of online disputes over intellectual property, online speech, and other issues. DMCA-like processes have now expanded to providers as diverse as advertising networks and payment systems. At the same time, section 512-style notice and takedown has been adopted in jurisdictions around the world. The basic process—though sometimes with variations that sharply tilt the balance among complainants, OSPs, and targets in different directions—has now been "woven into the fabric" of much of the Web world-wide. Tracing the tendrils of this reach illustrates that the notice and takedown process is here to stay, and suggests

---

45 *Viacom*, 676 F.3d at 30 (quoting Capitol Records, Inc. v. MP3tunes, LLC, 821 F. Supp. 2d at 635); *see also* Io Grp. v. Veoh Networks, 586 F. Supp. 2d 1132, 1151 (N.D. Cal. 2008) ("[T]he plain language of section 512(c) indicates that the pertinent inquiry is not whether Veoh has the right and ability to control it [sic] *system*, but rather, whether it has the right and ability to control the *infringing activity*.").

46 UMG Recordings, Inc. v. Shelter Capital Partners LLC, 718 F.3d 1006, 1029 (9th Cir. 2013) (Shelter Capital) (citing Viacom, 676 F.3d at 38 and quoting Capital Records, Inc. v. MP3tunes, LLC, 821 F.Supp. 2d 627, 635 (S.D.N.Y. Oct. 25, 2011)).

47 Shelter Capital, 718 F.3d 1006, 1029-30 (9th Cir. 2013).

48 The American Association of Independent Music (A2IM) et al., Response to Notice of Inquiry on "Copyright Policy, Creativity, and Innovation in the Internet Economy" at 7-8 (Dec. 10, 2010), http://www.ntia.doc.gov/files/ntia/comments/100910448-0448-01/attachments/Copyright%20NOI%20%28revised%29%20-%20121310%20%283334319%29.pdf (arguing that some court decisions have read the knowledge standard of § 512(c) too narrowly).

that our findings and recommendations may have application beyond the context of section 512 notice and takedown.

1. Repurposing Notice and Takedown for Other Problems and Extending it to Tertiary Providers Through Private Agreements

From their early days, the reach of the DMCA notice and takedown procedures began extending beyond their original mandate as Internet intermediaries adapted the required DMCA process to other problems. The notice and takedown procedures have proved to be a popular template for OSPs attempting to manage both adjacent content issues and the increasingly complex array of relationships between intermediary services. As OSPs implemented tools to manage DMCA notices, many adapted the model and infrastructure to trademark claims: Twitter, Facebook, and Yahoo! now process trademark claims through notice systems.[49] Others adopted some form of notice system to deal with other content issues, including the new "right to be forgotten" established in 2014 by the European Court of Justice.[50] As the online ecosystem grows more complex, notice practices have extended through private agreements to payment processors and ad networks—services outside the categories specified in the DMCA, and unlikely to incur secondary copyright liability.[51] These agreements and their "follow the money" strategy of addressing infringement have had strong While House support, via the office of the Intellectual Property Enforcement

---

[49] *See, e.g.*, *Trademark Report Form*, FACEBOOK, https://www.facebook.com/help/contact/284186058405647 (last visited Feb. 5, 2016); *Report a Trademark Issue*, TWITTER, https://support.twitter.com/forms/trademark (last visited Feb. 5, 2016); *Copyright and Intellectual Property Policy*, YAHOO, https://info.yahoo.com/copyright/us/details.html (last visited Feb. 5, 2016).

[50] EUROPEAN COMM'N, FACTSHEET ON THE "RIGHT TO BE FORGOTTEN" RULING (C-131/12) (2014), http://ec.europa.eu/justice/data-protection/files/factsheets/factsheet_data_protection_en.pdf; *Search Removal Request Under Data Protection Law in Europe*, GOOGLE, https://support.google.com/legal/contact/lr_eudpa?product=websearch (last visited Feb. 5, 2016).

[51] Following rightsholder negotiations with the payment industry, many of the largest payment processors—including American Express, Discover, MasterCard, PayPal, and Visa—developed voluntary best practices to withdraw payment services for sites selling counterfeit and pirated goods. *RogueBlock*, INT'L ANTICOUNTERFEITING COAL., http://www.iacc.org/online-initiatives/rogueblock (last visited Feb. 5, 2016). Administered by the International AntiCounterfeiting Coalition, a trade association representing brand owners, the "Portal Program" allows rights holders to report online sellers of counterfeit or pirated goods to credit card and payment processing networks. *See* KRISTINA MONTANARO, INT'L ANTICOUNTERFEITING COAL., IACC PAYMENT PROCESSOR PORTAL PROGRAM: FIRST YEAR STATISTICAL REVIEW 2 (2012), http://www.gacg.org/Content/Upload/MemberNewsDocs/October%202012%20Report%20to%20IPEC%20-%20FINAL.pdf. Since its launch in 2012, over 5,000 merchant accounts (representing over 200,000 websites) have been terminated through the program. *RogueBlock*, INT'L ANTICOUNTERFEITING COAL., http://www.iacc.org/online-initiatives/rogueblock (last visited Feb. 5, 2016).

Coordinator ("IPEC").[52] In the process, more types of intermediaries have been pulled into IP enforcement roles, structured by private agreements, "best practice" documents, and other norm-setting efforts.[53]

As the rightsholder groups work their way through the list of intermediaries that have some relationship to copyright infringing sites, the range of private agreements will surely grow. As with traditional notices, the integrity of the process turns on substantive review.[54]

2. The DMCA's International Reach

The reach of the section 512 model further extends through its replication or adaptation into numerous foreign copyright regimes. Since 1998, most countries have adopted intermediary copyright liability laws,[55] drawing either on the example of the DMCA or, after 2000, on the European Union's E-Commerce Directive, which implemented similar notice and takedown

---

[52] *See* Victoria Espinel, *Coming Together to Combat Online Piracy and Counterfeiting*, WHITE HOUSE BLOG (July 15, 2013, 8:33 AM), https://www.whitehouse.gov/blog/2013/07/15/coming-together-combat-online-piracy-and-counterfeiting. Advertising was the subject of a 2013 White House-brokered agreement between rightsholder groups and several of the major online ad networks, including Google, Yahoo, Microsoft and AOL. The resulting 'best practices' commit the partners to developing policies that "discourage or prevent… websites that are principally dedicated to selling counterfeit goods or engaging in copyright piracy and have no substantial non-infringing uses from participating in the Ad Network." BEST PRACTICE GUIDELINES FOR AD NETWORKS TO ADDRESS PIRACY AND COUNTERFEITING, http://www.2013ippractices.com/bestpracticesguidelinesforadnetworkstoaddresspiracyandcounterfeiting.html (last visited Feb. 5, 2016). Below this very loose level of commitment, the agreement recognizes a DMCA-like process of notices and counter notice as the basis for evidence collection and dispute resolution. In language that reaffirms the broad DMCA framework for exercising IP rights, the document specifies that:

> *"Rights holders are in the best position to identify and evaluate infringement of their intellectual property. Therefore, the Ad Networks agree that without specific, reliable notices from rights holders, Ad Networks lack the knowledge and capability to identify and address infringement." Id.*

Unlike the DMCA, the agreement is non-committal with respect to OSP follow up. A valid notice is supposed to trigger an investigation—not necessarily a strike or termination of service.

[53] Some commentators have criticized voluntary best practices on the basis that their effect is limited because they rarely involve the "worst" actors. *See, e.g.*, Bill Rosenblatt, *Ad Networks Adopt Notice-and-Takedown for Ads on Pirate Sites*, COPYRIGHT AND TECHNOLOGY (July 21, 2013), http://copyrightandtechnology.com/2013/07/21/ad-networks-adopt-notice-and-takedown-for-ads-on-pirate-sites/.

[54] Domain-name registrars, in particular, have come under pressure to make it more difficult for file-sharing sites to keep or create new domains. Andy, *Registrar Suspends Torrent Domain For DMCA Non-Compliance*, TORRENTFREAK (June 14, 2014), https://torrentfreak.com/registrar-suspends-torrent-domain-for-dmca-non-compliance-140614/; Andy, *Universal Lawyers: Registrar Liability in Torrent Case is "Common Sense,"* TORRENTFREAK (Feb. 8, 2014), https://torrentfreak.com/registrar-liability-in-torrent-case-is-common-sense-universal-lawyers-say-140208/. Other types of services are also being pulled into enforcement roles. For example, Cloudflare, a service that does not host websites or register their names but optimizes the speed of websites and offers distributed domain name server services, was recently ordered to stop providing services to a copycat version of Grooveshark, a streaming music website shut down by a copyright dispute. Andrew Chung, *Web Company Says Judge's Ruling Turns it into 'Copyright Police,'* REUTERS (June 4, 2015, 6:16 PM), http://www.reuters.com/article/2015/06/04/ip-fake-grooveshark-idUSL1N0YQ2L920150604.

[55] *See* Ignacio Garrote Fernández-Díez, *Comparative Analysis on National Approaches to the Liability of the Internet Intermediaries for Infringement of Copyright and Related Rights* (World Intellectual Prop. Org., Working Paper, 2011), http://www.wipo.int/export/sites/www/copyright/en/doc/liability_of_internet_intermediaries_garrote.pdf; Daniel Seng, *Comparative Analysis of the National Approaches to the Liability of Internet Intermediaries* (World Intellectual Prop. Org., Working Paper, 2011), http://www.wipo.int/export/sites/www/copyright/en/doc/liability_of_internet_intermediaries.pdf.

procedures and safe harbor protections.[56] In the wake of passage, the global standardization of DMCA-compatible rules also became a primary objective of US trade negotiators, who introduced intermediary liability requirements into a wide array of agreements in the 2000s.[57] Countries that adopted requirements through trade agreements include Chile, Singapore, Australia, Morocco, the Dominican Republic, Colombia, Panama, Peru, and South Korea.[58] Other countries, including India and South Africa, followed the US and EU models as they modernized their copyright laws.

The growing international agreement around basic principles of copyright intermediary liability nonetheless has an important caveat: the international system remains heterogeneous both with respect to the law and judicial interpretation of key activities. The EU's E-Commerce Directive, for example, does not require users targeted by takedown requests to be notified of requests, nor does it provide counter notice procedures—features that tilt the balance of expressive rights sharply toward rightsholders compared to the DMCA.[59] A few countries tilt the other way. Since 2010, Chile has required a court order for takedown requests.[60] Beginning in early 2015, Canada implemented a "notice and notice" system that requires only that Internet service providers and storage services[61] notify the target upon receipt of a notice, not that they take down content.[62] Highly fact-dependent debates about knowledge, inducement, financial benefit, and limitations and exceptions to copyright (such as fair use or fair dealing) are also embedded in distinctive national traditions of jurisprudence, with divergent outcomes.

As a matter of day-to-day practice, the rise of global U.S.-based OSPs has limited the role of this legal pluralism. Beyond its influence as a model, the DMCA also operates as *de facto* international law because the vast majority of notices are sent to US-based companies, which operate under it. Of the top ten global Internet destinations (in terms of unique monthly users), nine are US based.[63] Of the traffic to those sites, around 80% comes from outside the US.[64] The dominance of US companies in the ecosystem means that the different values that

---

[56] Council Directive 2000/31, 2000 O.J. (L 178) (EC). The E-Commerce Directive was largely inspired by the DMCA safe harbors, though it differs from the DMCA in several notable ways. For a comprehensive discussion, see Miquel Peguera, *The DMCA Safe Harbors and Their European Counterparts: A Comparative Analysis of Some Common Problems*, 32 COLUM. J.L. & ARTS 481 (2009).

[57] Seng, *supra* note 55, at 7.

[58] *Id.* China adopted similar principles in 2006. *Id.; see also* Lilian Edwards, Role and Responsibility of the Internet Intermediaries in the Field of Copyright and Related Rights (World Intellectual Prop. Org., Working Paper, 2011), http://www.wipo.int/export/sites/www/copyright/en/doc/role_and_responsibility_of_the_internet_intermediaries_final.pdf.

[59] *See* Council Directive 2000/31. However, several EU countries, including Finland, Hungary, Lithuania, Spain, and the UK, have introduced counter notice procedures.

[60] CTR. FOR DEMOCRACY & TECH., CHILE'S NOTICE-AND-TAKEDOWN SYSTEM FOR COPYRIGHT PROTECTION: AN ALTERNATIVE APPROACH (2012), https://www.cdt.org/files/pdfs/Chile-notice-takedown.pdf.

[61] Search engine providers are not required to forward a notice to the alleged infringer, but the safe harbor protections only provide protection to search providers' reproduction of work specified in the notice for thirty days after the notice is received. *See* GOV'T OF CANADA, BACKGROUNDER: NOTICE AND NOTICE REGIME (June 17, 2014), http://news.gc.ca/web/article-en.do?nid=858069.

[62] *Id.* The 'notice-and-notice' system is part of a new and, as yet, judicially untested system adopted as part of the 2012 Canadian Copyright Modernization Act. *See Id.*

[63] MARY MEEKER, INTERNET TRENDS 2014, at 130 (2014), http://www.kpcb.com/blog/2014-internet-trends (PowerPoint presentation reproducing January 2013 data from Comscore).

[64] *Id.*

shape legislative or judicial outcomes in Canada, or Germany, or Chile—such as different conclusions regarding knowledge or intent or the liability attached to linking, for example—have so far had limited purchase on the behavior of the core Internet intermediaries. Further, despite section 512's complex jurisprudence, several OSPs we interviewed as part of Study 1—including OSPs with European operations—described the DMCA as a force for stabilizing liability and safe harbor requirements relative to other less-developed doctrines, such as the E-Commerce Directive, where national courts have produced inconsistent interpretations of many of the key provisions, including those on matters as important as the scope of safe harbor protection.[65]

The situation is dynamic, however. United States intermediaries have been sued and held liable in other national jurisdictions over a variety of core business practices—defamation for search engine results in Italy,[66] online markets selling offensive memorabilia in France.[67] And some US-based companies operating internationally have developed mechanisms to remove content that violates local laws only on the local domain extensions where those laws apply.[68] However, the most potent effort to date to create a distinctive liability and takedown regime—the new "right to be forgotten" created by the European Court of Justice[69]—will test this method of compliance; as regulators seek to expand the reach of their authority, they have pushed for their decisions to apply beyond country-level domain extensions.[70]

---

[65] *See, e.g.*, Aleksandra Kuczerawy, *Intermediary Liability & Freedom of Expression: Recent Developments in the EU Notice & Action Initiative* 10 (Interdisciplinary Ctr. for Law & ICT, Working Paper 21, 2014), http://papers.ssrn.com/sol3/papers.cfm?abstract_id=2560257.

[66] On intermediary liability in Italy, see Nicolo Zingales and Eleonora Ortaglio, *ISP Liability in Italy* (Report to the International League of Competition Law Congress, 2013), http://ssrn.com/abstract=2338981.

[67] *See* Yahoo v. La Ligue Contre Le Racisme et L'antisemitisme, 433 F.3d 1199 (9th Cir. 2006).

[68] For example, Twitter's policy allows it to withhold a tweet from users in a specific country-level domain extension while keeping it available to users in the rest of the world. Twitter, *Tweets Still Must Flow*, Twitter Blog (Jan. 26, 2012), https://blog.twitter.com/2012/tweets-still-must-flow. Similarly, Google's Blogger service redirects blog readers to domain extensions specific to the reader's country, and removes content that violates local laws on just the country-specific domain extension to which the law applies—though it also provides a mechanism for readers who do not want to be automatically redirected to country-specific domains. *My Blog Redirects to a Country-Specific URL (ccTLD), Blogger Help*, Google https://support.google.com/blogger/answer/2402711?hl=en (last visited Feb. 5, 2016).

[69] *See* European Comm'n, *supra* note 50.

[70] For example, France's Commission Nationale de l'Informatique et des Libertés recently rejected Google's appeal from its order that Google remove requests results from all versions of its search engine—including Google.com—not just on the extension where a removal request came from, such as google.fr. *See, e.g.*, *Right to Delisting: Google Informal Appeal Rejected*, CNIL (Sept. 21, 2015), http://www.cnil.fr/english/news-and-events/news/article/right-to-delisting-google-informal-appeal-rejected/.

III. STUDY 1: OSP AND RIGHTSHOLDER ACCOUNTS OF NOTICE AND TAKEDOWN IN EVERYDAY PRACTICE

Study 1 is an in-depth qualitative account of online intermediaries' and rightsholders' practices and experiences with notice and takedown, based on close to three dozen survey responses and in-depth interviews with OSPs and major copyright holders. It provides a detailed picture of how copyright complaints are privately managed in light of rapid changes in technology and business practices.

Overall, we found that section 512 operates as the central framework under which OSPs and rightsholders alike handle infringement claims. OSPs all stressed the importance of the safe harbor to their ability to manage risk, and rightsholders all used section 512's takedown mechanisms extensively. At the same time, we found that notice-handling practices vary widely by sender and OSP. For some, the process works relatively well, and proceeds much as it has since the DMCA was implemented. For others, the notice and takedown procedure is fraying in the face of internet-scale infringement and automated notice sending; in some cases, it is being superseded by measures that go beyond and in some respects reverse the core responsibilities of the original framework, such as *ex ante* content filtering.

Following a discussion of methods, the study is divided into six sections:

A Fundamental Safe Harbor and a Split in Practice describes the central role that notice and takedown plays in managing online infringement, notwithstanding the deep divergence in OSPs' day-to-day experience and practice with notice and takedown. It describes the transition by some notice senders from human-mediated notice sending to practices that rely primarily on automated systems to detect infringing content and generate notices, the related rise in outsourcing enforcement to third-party rights enforcement organizations ("REOs"), and the measures affected OSPs have taken to manage the increased scale. It discusses how these systems are designed and used, and explains steps that notice senders and OSPs employ to reduce inadvertent targeting of non-infringing content.

Notice and Takedown in Operation describes notice handling at a variety of OSPs, including the many DMCA Classic OSPs that persist in hand reviewing notices. It looks at the emergence of OSP triage systems for processing notices. It describes particularly challenging aspects of takedown request processing, including difficulty locating infringing content with specificity, decision-making around boundary cases, and the problem of mistaken or spurious notices. It discusses the structural and practical limitations of counter notices and examines the implementation of repeat infringer policies. It concludes with a discussion of barriers to wide transparency reporting and public archiving of takedown requests by OSPs.

Divergence: the Move to DMCA Auto, DMCA Plus, and Beyond describes the shift by some OSPs away from DMCA Classic practice to large-scale, automated processing through "DMCA Auto" measures or through "DMCA Plus" measures that go beyond the statutory requirements. For those who have made them, these moves to DMCA Plus represent profound shifts in practice. This section also briefly discusses "Para DMCA" measures that fall outside of the section 512 framework, particularly site blocking and attempts to "follow the money" of infringement by putting pressure on advertising and other tertiary providers.

Compliance, Competition, and Market Power explores OSPs' concerns that DMCA Plus measures—which tend to be costly to implement and maintain—can provide a competitive advantage to well-resourced, incumbent OSPs, diminishing the pro-competition effects of the safe harbor.

The Role of Search explores the unusual place of search services in the notice and takedown ecosystem. Though search index links are unlikely to be infringing themselves, section 512 covers search providers, and rightsholders often send notices to search services as part of a wider strategy to limit the scale of infringement. This section examines, in particular, OSP perceptions of the disproportionate role played by Google Web Search in the enforcement ecosystem, both as a recipient of a large number of takedown requests and as an innovator in automated enforcement.

A brief **Discussion** section integrates and analyzes the Study 1 findings, and serves as a backdrop to Section V's analysis and recommendations.

A. METHODS

Study 1 uses qualitative interview and survey data to develop a descriptive approach to understanding the takedown ecosystem, focused less on law and policy than on the organization and practice of enforcement as described by our OSP and rightsholder respondents. It explores the experiences and practices of a range of stakeholders involved in online copyright disputes.

To develop as broad and accurate a picture of the notice and takedown process as possible, we employed mixed methods. We began with the center of the notice and takedown regime: the intermediary OSPs. The primary data are original, drawn from a detailed survey and a set of semi-structured interviews with 29 OSPs,[71] including platform, connectivity, and search providers. To this we added interviews with six major notice senders (including rightsholders, rightsholder groups, and REOs). These data were combined with publicly available information and secondary literature and reporting.

We reached the survey and interview group using a multi-layered selective sampling approach. We initially identified OSPs to contact using a list of the top-100 websites in the United States in terms of traffic[72] and then distributed the survey to these service providers' designated DMCA agents using the list of designated agents on the US Copyright Office website. Using the top-100 websites both limited researcher bias and ensured that initial respondents were

---

71 Respondents include video and music hosting services, search providers, file storage services, e-commerce sites, web hosts, connectivity services, and other user-generated content sites (e.g. blogs, social media, or other specialized platforms). Eleven of our OSP respondents' services are oriented towards individual users (e.g. user-generated content or entertainment services) and 18 are oriented towards both enterprises and individuals (e.g. cloud software and storage, productivity tools). 15 are small companies with fewer than 100 employees, six are medium companies with 100-499 employees, and eight are large companies with 500 or more employees.

72 ALEXA TOP SITES IN THE UNITED STATES, http://www.alexa.com/topsites/countries/US (last visited Feb. 5, 2016). Of the 100 emails sent to DMCA agents using the contact information provided on the list of designated agents on the US Copyright website, 18 of these emails failed to deliver (generating automated server error response).

likely to have sufficient traffic to engage with the notice and takedown process. We then employed a modified snowball approach. Survey recipients were encouraged to forward the survey to other potential respondents. The widely broadcast survey led to interviews, and interview respondents recommended other respondents. In order to maximize our chances of covering typical cases, we also approached OSPs who clearly represented a category or size of service not represented in our initial outreach. With regard to rightsholders, we focused on supplementary interviews with large rightsholders with strong interests in the operation of notice and takedown.[73] Some of the initial set of service providers and rightsholders then forwarded our request on to other potential participants. We took care to solicit participation from a wide range of stakeholders who were diverse in size, type of service, and position within the ecosystem. The result is not a statistically representative sample of the OSP or rightsholder communities, but it is a broadly inclusive one with more than one representative of nearly all of the major categories of service, as well as important differentiators within those categories.

We accepted no limitations on our reporting. No respondent reviewed this report prior to publication, except for a very small number of brief excerpts provided by us to request permission to quote a source without aggregating and anonymizing as described below.

The study has a number of limitations. We did not survey or interview targets of notices (those who are alleged to directly infringe), limiting our ability to report on this important stakeholder group; we did, however, ask our respondents about their interactions with targets. Both the OSPs and the senders, including REOs, with which we spoke tended to be professionalized, and all were reputable in their relevant industries. Our respondent pool did not include representatives of so-called "notorious" file sharing sites, nor did it include obviously abusive senders. The senders with which we spoke tended to be or represent large copyright owners with valuable properties, limiting our ability to comment on the interests of smaller copyright holders. We did not speak with rightsholders from the adult entertainment industry or their agents, a group that our respondents agreed are particularly prolific and aggressive users of the takedown system.

Our number of respondents, and the data we gathered, are also limited by the reluctance of a significant number of OSPs and senders to reveal some information about their practices—even with confidentiality protections in place. This tended to affect the level of detail about those practices that they felt that they could reveal. Some OSPs were not comfortable filling out the survey or speaking to us at all.[74]

---

73 These include trade associations, large organizational rightsholders, and REOs. They represent a range of the industry sectors that most use notice and takedown: games, movies, music, software, and the like.

74 There is also the possibility that respondents were not fully forthcoming or provided information with "spin" in their favor. Providing confidentiality helped limit this risk. Further, because most respondents revealed information that challenged public narratives around notice and takedown, including narratives that function in their interest, we think this problem was limited. There were occasional instances of respondents apparently repeating "talking points." These we discounted unless supported from other sources.

Data from the surveys and interviews are anonymized or aggregated, except where we received permission to report in greater detail. Other findings are drawn on publicly available information and data. As such, we stress that when reported experiences or descriptions identify an OSP or sender by name, this is not an indication that the identified example was a respondent.

Because of the sensitivity of many of these issues, our respondents overwhelmingly required assurances of confidentiality prior to participating in surveys or interviews. This prevents us from sharing original data with other researchers for independent review. Many OSPs hesitated to publicly reveal notice-handling practices due to the history of litigation in this area—in some cases based on concerns that they might attract the attention of third-party REOs. Rightsholders were concerned that revealing detailed information about specific enforcement practices could complicate their enforcement efforts or undermine competitive advantages. Because of these concerns, data from the surveys and interviews are anonymized or reported in general terms only, except where we requested and received permission to report in greater detail. Other findings draw on publicly available information and data. As such, we stress that when reported experiences or descriptions identify an OSP or sender by name, this *is not an indication that the identified example was a respondent*. The resulting narrative is at times very specific with regard to stakeholders and other times remains at the level of generalizations and categories.

B. A FUNDAMENTAL SAFE HARBOR, AND A SPLIT IN PRACTICE

One of the threshold questions for this study was whether notice and takedown is still relevant to online services, given how much they—and the practice of infringement—have both evolved since 1998. The answer to this question is a resounding "yes." Notwithstanding the serious shifts occurring for some stakeholders, and despite criticisms of the system by all involved, our respondents described notice and takedown—and especially, the compromises it strikes regarding OSP liability for copyright infringement—as constitutive of rightsholder and OSP responses to online infringement. Though different players experience notice and takedown in different ways, it remains foundational to how both OSPs and notice senders address copyright infringement and negotiate duties under the law.

The benefits of the safe harbor are, by all accounts, profoundly influential for all OSPs: As one OSP described it, OSPs "live or die" by section 512's safe harbors. As another put it, the DMCA takedown regime was a "godsend," allowing OSPs to operate in an increasingly complex online ecosystem in which connectivity services, platforms for content distribution and speech, search engines, and other services have widely differing relationships with their users and user infringement.

At the same time, our study shows that, in the last decade, a deep divergence in OSPs' day-to-day experience has appeared. OSPs split into three broad groups.[75]

[75] *See also* Annemarie Bridy, *Copyright's Digital Deputies: DMCA-Plus Enforcement by Internet Intermediaries*, RESEARCH HANDBOOK ON ELECTRONIC COMMERCE LAW (John A. Rothchild ed., forthcoming 2016), available at http://ssrn.com/abstract=2628827 (distinguishing two categories of DMCA-Plus enforcement practices: Type 1 DMCA-Plus entities, which operate under the DMCA but have adopted measures not required by the DMCA (generally corresponding to DMCA Plus OSPs in our analysis); and Type 2 DMCA-Plus entities, which would not generally attract secondary liability under U.S. law and do not operate under the DMCA statutory regime, but which have adopted measures beyond those required to obtain the safe harbor (generally corresponding to Para DMCA OSPs in our analysis)).

"DMCA Classic" OSPs: The first group contains OSPs for which the volume of takedown notices has remained relatively infrequent, and the traditional DMCA notice and takedown process, involving substantive human review of takedown notices, is still the norm. DMCA Classic OSPs commonly receive dozens or hundreds of notices annually, with minimal increases year-on-year, and tend to be operating outside of the areas where there has been heightened conflict over copyright (music, video, and web search). DMCA Classic OSPs comprised the majority of our Study 1 respondents. Although the Study was not able to survey the entire OSP eco-system to independently quantify its size, it appears that traditional notice and takedown continues to the dominant practice for the majority of OSPs.

"DMCA Auto" OSPs: The second group comprises OSPs that receive large numbers of notices generated by automated systems, which often rely on computer algorithms to both detect potential infringements and generate notices. DMCA Auto OSPs still operate according to their obligations under the statute—responding to takedown requests for specific alleged infringements that already exist on, or are linked to from, their systems—but on a massive scale. By comparison with DMCA Classic OSPs, DMCA Auto OSPs typically receive tens of thousands of notices (and in some cases millions) of notices per year, and have experienced steep increases in volume for each of the last several years. These OSPs' on-the-ground notice and takedown processes have significantly transformed in response to the large volume of notices they receive. DMCA Auto OSPs have had to shift to more automated notice-processing practices to handle the time and resource pressures of such large-scale notice processing ("DMCA Auto measures"), often sacrificing human review of the majority of automated notices they receive.

"DMCA Plus" OSPs: The third group comprises OSPs that have adopted measures that are not currently required by the DMCA safe harbor regime, but go beyond it. These include measures such as *ex-ante* filtering systems, hash-matching based "staydown" systems, direct back-end takedown privileges for trusted rightsholders, contractual side agreements with certain rightsholders that set forth additional protections and obligations for both parties, and other supplementary practices beyond the traditional notice and takedown regime ("DMCA Plus measures").[76]

<table>
<tr><th>DMCA CLASSIC</th><th>DMCA AUTO</th><th>DMCA PLUS</th></tr>
<tr><td colspan="2">Within DMCA secondary liability framework</td><td>Beyond DMCA Requirements</td></tr>
<tr><td rowspan="2">• OSPs receive relatively few notices
• OSPs engage in substantive human review of notices and respond to takedown requests with varying degrees of risk tolerance
• In many cases, OSPs' practices and concerns are not well-reflected in policy discussions</td><td colspan="2">• OSPs receive large numbers of automated notices, or feel vulnerable to a potentially large influx of notices or other forms of rightsholder pressure
• OSPs develop systems to triage notices (taking down content with little or no substantive human review)</td></tr>
<tr><td>• OSPs develop automated systems to process notices
• OSPs facilitate bulk sending and/or may prioritize processing of notices by developing "trusted" sender programs</td><td>• OSPs may give senders backend access with direct takedown privileges
• OSPs may manage content by integrating content filters or hash-based "staydown" systems
• OSPs and rightsholders may agree to side deals to supplement copyright enforcement under the DMCA framework</td></tr>
</table>

Table 1: Characteristics of DMCA Classic, DMCA Auto, and DMCA Plus OSPs

[76] This category is similar to Bridy's Type 1 DMCA Plus category. *See id.*

In addition to these three broad groups of OSPs, we also identified a final group that comprises service providers that fall beyond the section 512 framework, which likely would not incur secondary liability for user copyright infringement under U.S. law.[77] These "Para DMCA" OSPs may include advertising and payments providers, for example. Though we spoke directly only with OSPs that fall under the notice and takedown framework, enforcement efforts increasingly extend to these Para DMCA OSPs, and they often make up part of an enforcement ecosystem within which DMCA OSPs are acting.

Measured by the number of individual OSPs in each category, the DMCA Classic category—perhaps surprisingly—comfortably dominates the online ecosystem.

Measured by the number of individual OSPs in each category, the DMCA Classic category—perhaps surprisingly—comfortably dominates the online ecosystem. Other measures, such as market share or numbers of notices, would likely provide different pictures.[78] Section 512, however, regulates all covered entities, regardless of their size or other characteristics. Because of this, counting OSPs is the appropriate measure: reforms or changes in case law affect all covered entities, making it important to understand their potential effects at the entity level.

1. Stability Concerns

Some OSPs considered their status uncertain. The split into Classic, Auto and Plus is not wholly stable for two reasons. First, DMCA Classic OSPs described varying levels of concern that floods of automated notices or other pressure to implement DMCA Plus measures might imminently arrive. Many DMCA Classic OSPs did not see automation as an imminent problem. Other DMCA Classic OSPs, however, expressed concern that their situation is fragile. These worried that they differ from the major targeted services primarily in terms of lower visibility to rightsholders and REOs rather than any fundamentally different approach to user infringement. Since most of our respondents lacked awareness of other actors' practices and reasoning, some of these OSPs were unclear as to why they had not yet attracted floods of notices. They worried that, given the impact of automated systems, they were depending more on the inattention or forbearance of rightsholders and REOs rather than the protection of the law. These OSPs expressed concern that, despite being reputable services with legitimate business models not based on infringement, they, as one put it, had "flown under the radar" of senders' attention. OSPs in this group felt strongly that they were complying with the safe harbor's requirements, but that this was not necessarily enough to preclude either significant increases in the numbers of takedown notices sent to them or pressure to adopt expensive and aggressive automated enforcement systems in the future.

77 This category is similar to Bridy's Type 2 DMCA Plus category. *See id.*

78 Measured by market share or number of users, our data are unclear. Perhaps surprisingly, some DMCA Classic respondents were large entities with large numbers of users. Our three dozen respondents did not provide a broad enough picture to know whether market share or users makes much difference. Nor was type of service especially availing. Measured by the number of takedown requests across the online ecosystem, DMCA Auto and DMCA Plus are almost certainly dominant—the millions, tens of millions, or even more requests they receive overshadow the tens, hundreds, thousands (or, a few cases, more) requests DMCA Classic OSPs receive—but this simply reflects their position in the center of the policy and legal disputes over online copyright infringement and takedown. The reasons why they are there, and others are not, is not always clear.

Second, all DMCA Classic OSPs, regardless of whether they considered their situation to be "fragile" in terms of notice handling capacity, expressed concern that legal or policy changes aimed at resolving debates between DMCA Plus OSPs and major rightsholders could impose DMCA Auto or Plus requirements that they did not need, could not afford, and—in several cases—considered antithetical to their commitment to their users' expressive interests. They described being left aside in policy debates and news accounts skewed by attention to the relatively few DMCA Plus entities, and in some cases, being very limited in the resources they could devote to changing this. Several also voiced concern that, even in the absence of formal policy or new law, decisions by some high-profile OSPs to implement DMCA Plus measures could set norms that could push DMCA Classic OSPs to follow suit. These were concerned that norm shifting might eventually affect the standards for qualifying for the statutory safe harbor.

Beyond the basic split into DMCA Classic, DMCA Auto, and DMCA Plus practices, however, the potential "success" or "failure" of notice and takedown plays out daily in the diverse, and nuanced, approaches both OSPs and rightsholders take to implementing notice and takedown. We explore these practices in the remainder of Study 1.

2. Automation and the Professionalization of Notice Sending

Until recently, notice and takedown had an artisanal quality, conducted mostly by hand and with an element of human judgment on the part of both notice senders and receivers. The laboriousness of identifying infringing materials and generating takedown notices sharply constrained the system's overall scale. Between 2001 and 2008, Google received only 1,354 requests for the takedown of search results.[79] With the exception of ISPs providing connectivity services, which began receiving very large numbers of takedown requests in the early 2000s as part of rightsholder litigation to force disclosure of peer-to-peer user identities, every OSP practiced human review of incoming notices. This still holds true for a majority of our OSP respondents. For some, the scale is small: nine of our twenty-nine OSP respondents reported receiving fewer than 100 takedown notices per year in 2013.[80] Consistent with our findings, the handful of publicly available transparency reports provide examples of OSPs—some with significant web traffic—that receive small numbers of notices, that report rejection rates that indicate some substantive scrutiny of requests. For example, in 2014, Wikimedia Foundation (home of Wikipedia) reported receiving twenty-one takedown requests (and rejecting 76%)[81] and Reddit reported receiving 176 takedown requests (and rejecting 62%).[82]

By the early-to-mid 2000s, some rightsholder groups and their increasingly professionalized REO agents adopted software bots to crawl the web for infringing material. All of the major rightsholders and REOs with which we spoke described deploying automated systems that search sites for title matches, artist matches, or other indicators of unauthorized content. By design, such systems are intended to address large-scale Internet infringement by moving far beyond human capacity to search for and identify possible infringements. As these automated

---

[79] Seng, *supra* note 5, at 444.

[80] Seven of these nine OSPs reported that these notices included fewer than 100 individual takedown requests; the remaining two did not provide the number of takedown requests. Several other OSP respondents either declined to state or could not provide specific numbers, but described falling squarely in this small-scale processing category.

[81] *Requests for Content Alteration & Takedown*, WIKIMEDIA, *supra* note 12 (see "DMCA Requests, and How We Responded" subsection for details).

[82] REDDIT, REDDIT TRANSPARENCY REPORT, *supra* note 12.

systems proliferated, the number of notices sent to some major OSPs skyrocketed. For some OSPs, this led to a rapid, dramatic increase in the number of notices they receive. In 2009, Google received 4,275 notices.[83] In 2012, it received 441,370 notices.[84]

The growing number of notices tells only part of this story. Whereas DMCA Classic notices commonly request takedown for single alleged infringements, automated systems can pack hundreds, thousands, or—in some cases—tens of thousands of takedown requests into single submissions. In 2012, Google's 441,370 notices contained over 54 million individual takedown requests.[85] In 2013, the company processed over 230 million takedown requests.[86] In 2014, it processed 345 million.[87] As this report was being prepared for publication in early 2016, Google was receiving between 17 and 21 million requests *a week* for its Web Search service.[88]

Google Web Search is an outlier in the ecosystem: few OSPs (among those that disclosed information to us or report publicly) come close to receiving this number of takedown requests. Giganews—a Usenet provider—is one of them: it blocked access to more than 500 million individual messages for a one-year period between 2012-2013.[89] In talking with Usenet providers, we found that these very high numbers are in part due to the Usenet-specific fact that one infringing file may be broken into many individual Usenet messages, each of which must be individually identified for takedown in order for the Usenet provider to find it and respond.

However, although such high numbers are uncommon overall, some other OSPs have also experienced prodigious growth. For example, Twitter received 6,646 takedown requests in 2012.[90] In 2014, it received 25,847 takedown requests.[91] In our survey, one OSP respondent reported receiving 120,000 notices in 2012 and 220,000 in 2013. Another reported receiving 8,300 takedown notices in 2012, jumping to 23,500 in 2013. One of our respondent OSPs reported that circa 2009, it received fewer than ten notices per month on average. This respondent now receives over *a million* takedown requests per month. Both OSPs and rightsholders tended to describe this growth as linked to the falling cost and growing capacity of large stakeholders to produce and handle notices.

---

[83] Seng, *supra* note 5, at 444.

[84] *Id.*

[85] *Id.* at 460-461.

[86] *Hearing on Section 512 of Title 17 before the H. Judiciary Subcomm. on Courts, Intellectual Prop., & the Internet*, 113th Cong. 47 (2014) [hereinafter *Hearing on Section 512 of Title 17*] (testimony of Katherine Oyama, Senior Copyright Policy Counsel, Google Inc.).

[87] Joe Mullin, *Google Handled 345 Million Copyright Takedowns in 2014*, Ars Technica (Jan. 6, 2015, 1:05 PM), http://arstechnica.com/tech-policy/2015/01/google-handled-345-million-copyright-takedowns-in-2014/.

[88] *Transparency Report*, Google, *supra* note 12.

[89] Perfect 10, Inc. v. Giganews, Inc., No. CV 11-07098-AB SHX, 2014 WL 8628031, at *10 (C.D. Cal. Nov. 14, 2014) ("Indeed, pursuant to such requests [notices that included a Message-ID], Giganews blocked access to more than a half-a-billion individual messages between November 6, 2012 and November 6, 2013.").

[90] *Copyright Notices: DMCA Takedown Notices*, Twitter, https://transparency.twitter.com/copyright-notices/2015/jan-jun (last visited Feb. 5, 2016) (see 2012 tabs for Jan. 1–June 30 and July 1–Dec. 31).

[91] *Id.* (see 2014 tabs for Jan. 1–June 30 and July 1–Dec. 31).

Several rightsholder respondents also emphasized the growing technical sophistication of unauthorized file-sharing sites as a reason underlying the growth in notices and automated takedown techniques. They described effective takedown as more difficult now that some file-sharing sites have developed automated systems that facilitate the proliferation of content. For example, some popular files-sharing sites mirror content across multiple domains as an intentional strategy to complicate enforcement efforts. Other services, like BitTorrent, effectively mirror content, even if complicating enforcement efforts is not intended. Still other providers create copycat sites that mirror popular file-sharing sites to capitalize on their brand recognition. Direct download and streaming services (as distinct from peer-to-peer services) commonly separate linking and indexing from hosting functions—thereby adding one or more additional intermediaries between the user and the file—making targeted enforcement more difficult. Other services use automated systems that rotate links to underlying files, making URL-based takedowns a temporary remedy at best. Such automated strategies add to the more general resiliency of large file-sharing communities, whose members can quickly repost removed content. In this respect, the escalating use of automated systems to detect infringing material and send notices is an effort to compensate for increased opportunities for infringers to post content and the declining relevance of the unique 'link' in determining access to unauthorized files.

Sophisticated players' rapid adoption of automated systems is closely connected to the growing professionalization of large-scale enforcement, characterized by the emergence of specialized "content protection" teams in major trade associations and media companies, and by the growth of the REO sector that sells services to them. There has been very little research on this new tier of commercial players, but it is clear that professionalized enforcement is a crucial source of large-scale noticing. Urban and Quilter noted the emergence of REOs sending notices to a Texas ISP between 2004 and 2007. Seng's 2014 review of notices from the Lumen archive traced a further shift toward REO sending. Between 2008 and 2012, Seng found that REOs issued between 37% and 60% of notices sent to Google annually.[92] Of the top twenty takedown request submitters to Google Web Search in 2015, thirteen were REOs, four were record, film, or software trade associations, and three were individual studios.[93] Our interviews with OSPs broadly confirmed this. Those OSPs that received large-scale, automated notices described them as emanating from professionalized enforcement groups.[94]

In the course of this professionalization, the practices of small and large rightsholders have diverged. Some individual rightsholders, in particular, have publicly expressed their frustration with the resource-intensive nature of detecting infringing content in the absence of automated systems.[95] In absolute terms, individuals' use of the system appears to be substantial and growing. Between 2008 and 2012, individuals as a collective group were the fifth largest sender

---

[92] Seng, *supra* note 5, at 396.

[93] *Transparency Report*, Google, *supra* note 12 (see the *Reporting Organizations* subsection for most recent data).

[94] See *infra* Section IV.B.1.a.

[95] *See, e.g.*, *Hearing on Section 512 of Title 17*, *supra* note 86, at 54 (2014) (statement of Maria Schneider, Grammy Award Winning Composer/Conductor/Producer, Member of the Board of Governors, New York Chapter of the Recording Academy) (stating that she "must spend countless hours trying to take [her illegally uploaded music] down, mostly unsuccessfully."). Interestingly, however, our examination of Google Image Search notices in Study 3 indicates that at least some smaller senders have developed methods that allow them to generate many notices, although as a statistical matter, the quality of these notices was relatively low. *See infra* Section IV.C. These findings suggest that information-sharing and education efforts might help smaller senders to at least some degree. *See infra* Section V.D.

of notices submitted to the Lumen archive.[96] And as we describe in Study 3 below,[97] individuals and small businesses sent most of the notices to Google Image Search. But the percentage of their contribution to the overall number of notices sent to Google services, and especially of individual takedown requests, has plummeted. Between 2002 and 2005, individuals accounted for around 18% of the notices sent to Google Web Search (a total of 92 notices).[98] By 2012, the number of notices had climbed to 9,425, but this represented only 2% of the total number of takedown notices.[99] The 9,425 notices contained 120,000 takedown requests, representing only .2% of the 54 million total requests.[100] Some representatives of smaller senders considered themselves relatively disadvantaged by limited access to automated methods.[101] As we discuss in Study 3, the extent of this disadvantage is unclear—certainly some individual senders and small businesses are capable of sending large numbers of notices.

a. Design and Use of Automated Detection Systems

The rising use of automated systems to detect infringing content prompts concerns about their accuracy. A process that relies on machine judgment for detection and that readily scales to sending millions of notices has the potential to make mistakes on a similarly large scale. In interviews, rightsholders that employ these systems expressed sensitivity to this issue. They stressed that the automated systems they use employ measures that attempt to narrowly target content and limit mistakes. The techniques described to us were generally relatively simple. For example, rightsholders described limiting the terms used by web crawlers for title- or artist-matching in order to avoid targeting non-infringing material such as mash-ups, parodies, reviews of music or movies, and other non-infringing "dolphins" that might be caught in the net.[102]

Most major senders also described supplementing their automated systems with limited manual procedures and triggers for manual, human review to reduce the likelihood of misidentifying infringing content. For example, one sender

Several rightsholders described treating sites "dedicated to infringement" differently from more general-purpose sites when training and using automated notice systems. One sender, for example, gave less scrutiny to machine-identified infringement on sites "trading in copyrighted content" than machine-identified infringement on fan sites and others with a "community" character.

96 Seng, *supra* note 5, at 448.

97 *See infra* Section IV.C.

98 Urban & Quilter, *Efficient Process*, *supra* note 5, at 652.

99 Seng, *supra* note 5, at 412. To identify individual senders, Seng assumed that reporters with redacted information or identified as "redacted" or "private" are individuals rather than corporations. *Id.* at 448.

100 *Id.* at 412. Available data shows some variation across types of service. In an unpublished 2009 study of notices sent to a major Texas ISP during 2006, Urban and Quilter found that REOs and trade associations sent 94% of the 512(a) notices but only 46% of 512(c) notices (where the ISP also acted as a content host). Jennifer Urban & Laura Quilter, Undue Process: Challenges for Rightsholders and Service Providers in Implementing Section 512's Notice and Takedown Provisions 12, 17 (Jan. 2009) (unpublished conference manuscript) (on file with authors).

101 *See, e.g.*, *Hearing on Section 512 of Title 17*, *supra* note 86, at 54 (2014) (statement of Maria Schneider, Grammy Award Winning Composer/Conductor/Producer, Member of the Board of Governors, New York Chapter of the Recording Academy) (stating that she "must spend countless hours trying to take [her illegally uploaded music] down, mostly unsuccessfully.").

102 We discussed some of these methods with rightsholders in more detail, but they understandably asked us to refrain from describing specific methods that might allow file-sharing sites to develop counter measures.

distinguished between the techniques it uses to manage infringing content on websites that are "trading in copyrighted content" from those that have a "community" character, such as fan sites. This sender described using initial or periodic human review of sites detected by its automated infringement crawler to determine they are clearly "dedicated to infringement." Content that is identified by automated systems as infringing receives different treatment depending on how the site is manually classified by the human reviewer. Material on sites considered "dedicated to infringement" receives less scrutiny before a takedown request is sent than does material on "community" sites, which will receive additional review. This sender also conducts periodic reviews to see if a site is changing its behavior and adjusts its response accordingly.

Other rightsholders described similar triage approaches that attempt, on the one hand, to more efficiently flag infringements, and on the other, to avoid alienating user communities and fan bases. Some described using extensive manual review for edge-case uses that they might consider technically infringing but ultimately tolerated, such as fan work.

Consistently, rightsholders stressed the importance of conducting human cross-checks on automated results to guard against systemic inaccuracies, like targeting previously removed material or non-existent pages, and against collateral damage, like targeting legitimate content by requesting removal of too much of a page or site.

Rightsholders stressed that responsible senders conduct human cross-checks on automated results to guard against systemic inaccuracies.

DMCA Auto and DMCA Plus OSPs also described human attention as important to limiting problems with automated decision-making. OSPs described a variety of triaging systems that escalate certain notices for human review, sometimes through several layers of increasingly expert review. One OSP's triage system ends in a biweekly meeting in which high-level employees decide whether, as the respondent put it, to "bet the company" by deciding not to remove wrongly targeted material.

Both rightsholders and OSPs acknowledged that automated systems, even if responsibly deployed, have limited capacity to avoid mistakes.

Still, both rightsholders and OSPs acknowledged that in an automated environment, most decisions are made without human intervention by either the sender or the recipient OSP. And automated systems, even if responsibly deployed, have limited capacity to avoid mistakes. Some OSP respondents expressed concern that these systems are particularly ill-suited for complex legal decision-making, such as assessments of whether a particular use may be making a fair use of copyrighted content. Sender respondents acknowledged flaws in automated systems, and, as described above, take steps to avoid the misidentification of targeted material, but accepted some inaccuracy as the cost of mass enforcement. Unsurprisingly, OSPs and senders tended to disagree on the question of relative responsibility for evaluating infringements and assessing claims. OSPs described discomfort with both the cost and the substantive judgment associated with assessing the accuracy of takedown requests, and described the counter notice procedure—discussed further below—as inadequate. Senders, however, tended to describe counter notices as a meaningful, if uncommonly used, backstop against mistakenly targeted content.

C. NOTICE AND TAKEDOWN IN OPERATION

Notice handling practices vary widely among OSPs, depending in part on their exposure to automated notices and their choices about how to handle them. Unsurprisingly, costs and resource allocation for notice handling also varies widely among OSPs. The largest team reported thirty full-time positions dedicated to reviewing takedown notices. This appears to be an unusually high number, though we cannot say how unusual, as several large recipients declined to provide details about their staffing. Most OSP respondents received relatively fewer numbers of notices, and consequently reported smaller staffs. Most reported having between one and three full-time positions dedicated to reviewing notices, generally as part of teams dealing with wider compliance issues around privacy, user conflicts, trademark, and other issues. In several cases—including for some services with very large user communities—notices are infrequent enough to require less than one full-time position.

Notwithstanding these differences, some relatively consistent notice handling practices emerged from our interviews. Most OSPs that receive more than a handful of notices implement triaging systems to simplify notice workflow. OSPs reported surprising consistency in the types of requests that they are most comfortable rejecting and the categories of decisions that require the most resources to address. Within the category of notices that require more review, OSPs' tolerance for risk and stance on the importance of user rights drive takedown decisions.

1. Statutorily Required Statements and the Emergence of Web-Forms

Notices that are, as one OSP put it, "defective on their face"—before even getting to questions about the underlying legal claims—are primary targets of OSP triage systems. Section 512 provides a relatively clear set of required elements for a valid notice, with which senders must "substantially" comply.[103] In particular, section 512 establishes certain requirements that mirror fundamental due process concepts: a properly actionable request (e.g., the complaint relates to copyright, not some other issue such as privacy, defamation, or trademark); a properly identified rightsholder or agent; an accompanying sworn statement of accuracy; and specific and actionable identification of the alleged infringing material and its location. Nearly all OSP respondents described taking steps to catch notices missing readily verifiable elements of the statutory requirements, for example: a signature; a statement of good faith; a statement that the notice is accurate. This is true regardless of whether an OSP uses an automated web-based system that captures these requirements or whether it receives and reviews notices by hand.

Prior to 2009, OSPs accepted notices primarily via email, with additional provisions for postal mail and fax. Starting with Google in 2009, a few large OSPs began to develop

[103] 17 U.S.C. § 512(c)(3)(A) sets out the elements of a notice that are required for it to be effective. 17 U.S.C. § 512(c)(3)(A) (2012). The notice must include (i) a physical or electronic signature of the copyright owner or agent; (ii) identification or the copyrighted work or a representative list of list of copyrighted works; (iii) identification of the allegedly infringing material and information that is reasonably sufficient to permit the OSP to locate the material; (iv) the complainant's contact information; (v) a statement that the complainant has a good faith belief that the use of the material is not authorized; and (vi) a statement that the use of the notification is accurate and, under penalty of perjury, that the complainant is authorized to act on behalf of the rightsholder. § 512(c)(3)(A)(i)-(vi).

online forms to standardize notice submission and expedite their handling.[104] Currently, web forms are slowly becoming more common for both large and small OSPs. Many OSP respondents described gradually formalizing their notice submission process over the past five or six years, with some specifically moving to web-based forms as the primary mechanism for notice intake.[105]

From the OSPs' perspective, form-based systems have a number of benefits. First, they can "pre-code" some of the statutory requirements by prompting senders to provide statements and other information required by the statute, such as contact details and a signature.[106] This simplifies the process of screening for defective notices: the sender either fills in the required fields or does not.[107] Second, web forms facilitate following up with notice senders to complete or clarify a request—a process that several DMCA Classic OSPs indicated had previously consumed a disproportionate amount of staff time.[108] One OSP described this as a "customer service issue," while others pointed to the DMCA's statutory requirement for follow-up where the submitter "substantially" complied with several specified requirements but missed others.[109]

In some cases, forms help OSPs manage their user populations and educate them about the proper bases for complaints. OSPs reported that open-ended notice processes can produce considerable confusion on the part of non-IP professionals, leading small-scale senders in particular to submit improper notices. These notices may, for example, attempt to shoehorn trademark, privacy, defamation, censorship or other claims into DMCA procedures.[110] Web forms can help clarify appropriate subject matter, and channel or discourage improper complaints. OSPs that regularly manage intra-community disputes (as in the case of many sites focused on user-generated content) also reported that web forms, together with educational materials, could help to guide complainants away from making copyright complaints over unrelated disputes and toward channels where their complaints can properly be addressed. For example, one OSP

OSPs that mediate intra-community user disputes report that web forms can increase the overall quality of complaints and, together with user education on copyright, can reduce bickering among users.

[104] Google introduced an online form for Blogger notices in 2009 and for most of its other services, including WebSearch, in 2011. *See* Jonathan Bailey, *Google Accepts Online DMCAs for Blogger*, PLAGIARISM TODAY (April 14, 2009), http://www.plagiarismtoday.com/2009/04/14/google-accepts-online-dmcas-for-blogger/; Jonathan Bailey, *Google Accepts Form DMCA Notices for All Services*, PLAGIARISM TODAY (March 30, 2011), http://www.plagiarismtoday.com/2011/03/30/google-accepts-form-dmca-notices-for-all-services/.

[105] OSPs have also built web-based forms to handle other categories complaints, resulting in separate tracks for copyright, trademark, and other types of complaints against users.

[106] For examples of OSP webforms, see *infra* note 117.

[107] Of course, this does not necessarily solve problems with the underlying accuracy of sender contact information or sworn statements, but it does avoid problems with senders not knowing to provide the information.

[108] The reason for this laborious practice is that an incomplete notice may trigger actual or red flag knowledge of infringement. § 512(c)(3)(B)(ii).

[109] Where an OSP receives a notice that does not comply with the requirements of 17 U.S.C. § 512(c)(3)(A)(i)-(vi), but substantially complies with (ii), (iii), and (iv), the notification shall not be deemed to trigger knowledge if the OSP promptly attempts to contact the complainant to assist in the receipt of a notification that substantially complies with all of the provisions. § 512(c)(3)(B)(ii).

[110] *See* Urban & Quilter, Undue Process, *supra* note 100, at 21-22 (describing such confusion on the part of senders).

described having to manage frequent community "bitching sessions" regarding conflicts over user behavior that was not copyright infringement. For this OSP, implementing a form-based system, together with educational materials on the site about what constituted a valid copyright complaint, significantly reduced the frequency with which "bitching sessions" became formal complaints to the OSP. Overall, OSPs reported that forms tend to increase the overall quality of complaints, and in some cases, reduce their number.

OSPs that do not use web forms expressed concern that web forms could expose the service to automated notice senders.

Given these virtues, turning to form-based submission systems may seem to be an obvious choice. But not all OSPs agree. OSP respondents that do not employ web forms expressed concern that web forms could create new problems by lowering the cost of submitting of notices to them and—especially—by exposing the service to automated notice senders. These OSPs expressed concern that a web form could encourage floods of notices, perhaps of low quality, that they lacked resources to handle. Google's experience is the universal reference point for these concerns. Several OSPs viewed the increase of notices to Google as at least partly related to its introduction of a web form that could be used by automated systems. Our data is inconclusive about whether this is a genuine risk. A few OSPs reported an increase in the number of notices received following the introduction of a web form, though in most cases, the increase could not definitely be linked to the web form. Others that adopted forms reported no significant post-adoption change in the numbers of notices received. A few use CAPTCHAs and other techniques to try to meter automated sending. We do not see a pattern that can be readily disentangled from the general rise in notice sending, or from the differences in rightsholder attention to different types of services.

2. Locating Infringing Content: A Complicating Factor

Section 512 requires that a notice specify the location of allegedly infringing material.[111] OSPs consistently reported that this is the most common weak point in section 512's chain of requirements, one that is not easily remedied by shifting to web forms. For several OSPs, attempting to identify allegedly infringing material based on imprecise location pointers represents the most challenging and resource-intensive aspect of takedown. Rightsholders may not identify each specific URL or other identifier where a work is found, instead specifying titles, artists, entire search result pages, or similar identifiers. In response to imprecise requests, OSPs may struggle, or find it impossible, to locate the allegedly infringing material on their site or to know whether every instantiation of a title, artist, or other identifier on an identified page is truly infringing. The DMCA provides some protection to OSPs in this regard—notices that do not identify the specific location of the alleged infringement are not sufficient to confer knowledge on the service provider[112]—but the high potential cost of legal challenges leads OSPs to try to identify the material when possible.

OSPs reported that the most common weak point in notices is identifying the location of allegedly infringing material clearly enough that OSPs can take action.

[111] § 512(c)(3)(A)(iii).

[112] *See, e.g.*, Perfect 10, Inc. v. CCBill, LLC, 488 F.3d 1102, 1114 (9th Cir. 2007); UMG Recordings, Inc. v. Veoh Networks Inc., 665 F. Supp. 2d 1099, 1108-09 (C.D. Cal. 2009), *aff'd sub nom.* UMG Recordings, Inc. v. Shelter Capital Partners LLC, 667 F.3d 1022 (9th Cir. 2011), *opinion withdrawn and superseded on reh'g*, 718 F.3d 1006 (9th Cir. 2013), *and aff'd.* 718 F.3d 1006 (9th Cir. 2013); Hendrickson v. eBay, Inc., 165 F. Supp. 2d 1082, 1093 (C.D. Cal. 2001).

In practice, many takedown requests point to links which may be at several removes from the file in question. Increasingly, both "pirate" and legitimate sites are dynamic, complex assemblages of content drawn from multiple sources, making precise identification of the source difficult for rightsholders and increasing the potential for ripple effects in the event of an imprecise takedown. System architecture can also contribute to the complexity in identifying infringing material. For example, Usenet services—which have become a prominent target of notice senders in the last few years—are built on a protocol that requires breaking up large files across multiple "messages," requiring a multi-part process of identification for the OSP to fully remove a single targeted work.

OSPs described increased difficulties when takedown requests target pages that may also include non-infringing content. Common examples include requests to remove search result pages that may include both infringing and non-infringing content, requests to remove comment threads that may contain an infringing link somewhere in the thread, and requests to remove pages with dynamic content that may no longer contain the material in question. At the limit, OSPs reported that takedown requests can become *de facto* takedowns of whole sites, either through the volume of requests or the targeting of top-level pages.[113]

Most OSPs reported acting conservatively, taking down content in order to avoid liability even if it means also removing non-infringing content on the targeted page. On the other hand, some OSPs were able to develop policies that avoid the most sweeping requests. Search service respondents, for example, generally reported that they reject takedown requests targeting home page URLs. At the furthest extreme, OSPs reported receiving notices for content that was not on, or even linked to from, their systems.

One OSP that receives a large volume of notices reported that its staff invests a great deal of time and effort evaluating notices that eventually prove to be "false positives." This respondent explained that, out of fear of failing to remove infringing material, and motivated by the threat of statutory damages, its staff will take "six passes to try to find the [identified content]." In the end, the OSP stated that 7 to 8% of the takedown requests it receives refer to material that is not on its servers. Another respondent echoed this, stating that some senders send takedown requests for material that was removed a year or two before the notice was sent.

One respondent explained that, out of fear of failing to remove infringing material, and motivated by the threat of statutory damages, its staff will take "six passes to try to find the [identified content]."

The limited publicly available information confirms this phenomenon. Google's Transparency Report indicates when a takedown request is not addressed because the URL specified is a duplicate of a URL from a previous request.[114] Similarly, in our quantitative review of notices to Lumen, we found that some automated senders continued to target defunct sites.[115]

[113] This issue is not new and has continued throughout notice and takedown's tenure. Google began sending takedown notices it received to Chilling Effects (now Lumen) in 2002 after it received criticism for removing the top-level domain xenu.com (a site critical of Scientology) from its search index in response to a notice from the Church of Scientology. Gallagher, *supra* note 11. We heard of similar requests sent to other OSPs in our interviews.

[114] *Transparency Report*, GOOGLE, *supra* note 12 (see the FAQ subsection).

[115] *See infra* Section IV.B.2.a.

Although most responsible rightsholders make attempts to avoid it, duplicate notices do make it through to OSPs.

3. Evaluating the Substance of Claims

All of our OSP respondents, regardless of their size, the size of their teams dedicated to reviewing notices, or the number of takedown requests that they receive, described devoting the majority of their notice-processing capacity to a relatively small number of challenging notices: those that require considering the accuracy of the stated claims; possible fair use defenses; possible public domain scenarios; or other judgments about the limits of protected speech and expression. Such decisions are resource-intensive; one large OSP respondent reported that a difficult takedown request might be examined by as many as fifteen people before action is taken.

Nearly every OSP and several rightsholders expressed concern about the takedown of non-infringing content. In the telling, such cases are not uncommon. Nearly every OSP recounted stories of deliberate gaming of the DMCA takedown process, including to harass competitors, to resolve personal disputes, to silence critics, or to threaten the OSP or damage its relationship with its users. Although the proportion of problematic requests varied by type of OSP, every OSP told stories of takedowns that ignored fair use defenses or that targeted non-infringing material. Several echoed one respondent's view that "many copyright complaints… would obviously qualify as fair use; others are complete fabrications to remove content considered undesirable to the filer."

OSPs devote the majority of their notice-processing resources to managing substantively challenging notices. Several echoed the view that "many copyright complaints… would obviously qualify as fair use; others are complete fabrications to remove content considered undesirable to the filer."

Several respondents said that the most consistent predictor of a low-quality notice was whether it came from a first-time, one-off, or low-volume sender. OSPs noted that in their experience, these "small senders" are most likely to misunderstand the notice and takedown process, mistake the statutory requirements, or use it for clearly improper purposes. Accordingly, OSPs typically give notices from small senders more scrutiny prior to takedown. Several OSPs said that notices sent by small senders always trigger human review. One OSP reported that when its review team receives a notice from a first-time sender, the reviewing staff member will do a "gut check" and might do further research to verify whether, for example, the listed sending agent even exists. In some cases, this research reveals a fraudulent notice and the OSP will not take the content down. Another OSP described a similar experience and noted that it pays special attention to a takedown request when "something seems off with a notice," such as when the notice contains misspelled words or does not "seem particularly well-thought out." This respondent noted that notices sent by professional content companies do not present these flags.

Several respondents said that the most consistent predictor of a low-quality notice was whether it came from a first-time, one-off, or low-volume sender.

While all OSPs raised concerns about mistaken or spurious notices, they varied both in the strength of their concerns and the steps they take to identify and address the issue. Typically, these varied by type of service. Few OSPs have the staff or expertise to deal with large numbers of boundary cases. Most OSPs reported acting conservatively in order to avoid liability, opting to take down content even when they are uncertain about the strength of the underlying claim. In some cases, OSPs simply take substantive claims at face value and remove everything if notices conform to section 512's basic requirements. Four services among our respondents reported a takedown rate of 100%. OSPs uniformly described their conservatism as a result of necessarily prioritizing avoiding liability over taking risks that might protect expression. As one described it, "the process forces you to try to stay out of making judgment calls [and] to take [takedown requests] at face value."[116]

As one OSP described the priority on avoiding liability, "the process forces you to try to stay out of making judgment calls [and] to take [takedown requests] at face value."

In general, UGC services with manageable numbers of notices and loyalty-based communities expressed more willingness to push back on or reject requests. These OSPs tended to emphasize their concerns about the negative effect of mistakenly targeted content on users. Correspondingly, these OSPs expressed more willingness to accept the risk of declining to act on takedown requests, or in some cases, willingness to take the extra step of reaching out to notice senders to ask them to review or rescind the notice. For example, one OSP with a small team that hand-reviews each notice stated that it does not feel judgments about fair use should rest with the OSP. However, this OSP's staff will sometimes contact the sender to explain that the targeted material appears to be a strong candidate for fair use and to request that the sender rescind the request. The OSP explained that it takes the time to engage with senders of borderline notices because it "wants to be a fine upstanding member of the Internet community." It is occasionally successful at persuading the senders to withdraw requests. However, if a sender refuses to withdraw a request, the OSP will remove the item rather than take on liability risk itself.

One OSP explained that it takes the time to engage with senders of borderline notices because it "wants to be a fine upstanding member of the Internet community."

Rightsholders, too, expressed concern about the potential that they could mistakenly target content, both because of the danger to freedom of expression and, more pragmatically, because of the risk of public relations blowback or alienating a fan-base. Rightsholders with this concern described being likely to respond positively when mistakes are pointed out to them—if the mistakes are found. This concern appears to be particularly salient when enforcement passes through third-party REOs, which distance the rightsholder from situations that might favor tolerance. At times, OSPs noted, REOs appear to take actions with which rightsholders would not agree. In one case, an OSP that provides a UGC platform reported bringing to the attention of the rightsholder company over-aggressive targeting of fan work by a third-party REO employed by the rightsholder. In response, the rightsholder's international corporate headquarters promptly approved the use and withdrew the notice.

[116] OSPs varied in their sense of what constitutes "conservative" practice, however, and sometimes described quite sophisticated decision making. For example, one OSP hesitates to reject a notice based on a possible fair use defense for fear of leaving itself without the benefit of the safe harbor but nonetheless described a relatively detailed and nuanced handling of notices, explaining that it, for example, denies takedown requests where a sender claims to have copyrighted an idea. (Under U.S. copyright law, ideas are not protectable subject matter. 17 U.S.C. § 102(b) (2012)).

Our interviews with other rightsholders suggest that those with consumer-focused products might respond similarly, but it is less clear how often issues would be identified. The OSP in question receives relatively few notices overall and happened to have an unusually legally sophisticated reviewer who devoted the time and effort to track down the appropriate decision maker at the rightsholder company.

Occasionally, a rightsholder may independently elect to rescind the notice and ask the OSP to reinstate the content. The usual driver in such cases is bad publicity. In an example that is likely typical, one OSP reported receiving a query from a rightsholder asking why an item was removed after the target of a takedown request complained on social media the takedown was "unfair." As with OSP-identified problems, it is typical in such cases for the problematic notice to be sent by a REO. In these instances, the OSP will ask the REO sender to rescind the takedown notice and, if done, will reinstate the content.

As one OSP put it: it is "way too easy for spurious takedown notices to be filed," whether by individuals or by large automated systems sending tens or hundreds of thousands of requests.

There was almost universal agreement among OSPs that a lack of effective disincentives or remedies for erroneous notices amplifies the problem of mistaken or spurious notices. As one OSP put it: it is "way too easy for spurious takedown notices to be filed," whether by individuals or by large automated systems sending tens or hundreds of thousands of requests. While responsible rightsholders do take precautions to prevent mistakenly targeting content, there are limited legal incentives for them to do so. Some of the major OSPs ask notice senders to evaluate possible fair use or other exceptions when submitting a notice.[117] Others mention or even require rightsholders to check a box acknowledging 512(f) liability for knowing material misrepresentation.[118] The recent decision in *Lenz v. Universal* that copyright holders must consider fair use before sending a notice lends some legal support to these practices.[119] However, in other cases, neither the DMCA's requirement that the copyright owner must have a good faith belief that the use of the material is unauthorized nor its prohibitions against material misrepresentations have been held to require robust investigations by rightsholders.[120] To date, efforts to get

---

[117] *See, e.g.*, *Digital Millennium Copyright Act (DMCA) Notice*, AUTOMATTIC, http://automattic.com/dmca-notice/ (last visited Feb. 5, 2016); *Removing Content from Google*, GOOGLE, https://support.google.com/legal/troubleshooter/1114905?hl=en (last visited Feb. 5, 2016); *Notices of Infringement*, MICROSOFT, https://www.microsoft.com/info/FormForCloud.aspx (last visited Feb. 5, 2016); *Copyright Infringement Notification*, PINTEREST, http://www.pinterest.com/about/copyright/dmca (last visited Feb. 5, 2016); *Report Copyright Infringement*, TWITTER, https://support.twitter.com/forms/dmca (last visited Feb. 5, 2016).

[118] *See, e.g.*, *Digital Millennium Copyright Act (DMCA) Notice*, AUTOMATTIC, *supra* note 117; *Notification of Claimed Infringement*, DROPBOX, https://www.dropbox.com/copyright_complaint (last visited Feb. 5, 2016); *Removing Content from Google*, GOOGLE, *supra* note 117; *Notices of Infringement*, MICROSOFT, *supra* note 117; *Report Copyright Infringement*, TWITTER, *supra* note 117.

[119] Lenz v. Universal Music Corp., 2016 U.S. App. LEXIS 5025, at *16 (9th Cir. Mar. 17, 2016) (holding that fair use is "authorized by law" and a copyright holder must consider the existence of fair use before sending a takedown notification under § 512(c)). However, the court specifically limited this requirement to the sender's subjective belief rather than an objective determination. *Id.* at *17.

[120] One court described the "good faith" standard as a subjective standard that does not require a full investigation to determine the accuracy of the claim. Rossi v. Motion Picture Ass'n of Am., Inc., 391 F.3d 1000, 1003-4 (9th Cir. 2004); *see also* Tuteur v. Crosley-Corcoran, 961 F. Supp. 2d 333, 343-44 (Mass. 2013) (holding that "in enacting the DMCA, Congress did not require that a notice-giver verify that he or she had explored an alleged infringer's possible affirmative defenses prior to acting, only that she affirm a good faith belief that the copyrighted material is being used without her or her agent's permission.").

rightsholders to evaluate claims before sending function mainly as educational efforts aimed at small senders.

Further, although they could theoretically recover damages against rightsholders who make material misrepresentations, OSPs generally consider such remedies to be largely out of reach due to expense and slowness of court cases, the high standard of proof, and the fact that significant cost recovery is unlikely.[121] In contrast to the statutory penalties for infringement, which can run to $150,000 per infringed work, fraudulent or abusive takedown incurs only proven damages. Several OSPs expressed reluctance to pursue such claims, arguing that the standard of proof required for recovery is simply too high to give them sufficient expectations of success to justify a suit. Taking down is also much safer than leaving material up: it eliminates the possibility of suit against the OSP by the copyright holder. In theory, the target could sue the OSP for removing the material, but the target would also have to meet a high standard to recover, meaning that suits from that direction are highly unlikely.[122] One service provider stated that there is "no choice" but to take down content unless the notice is deficient on its face or obviously fraudulent. Another provider with a rigorous notice review board to review borderline cases described it as "betting the company" every time they decide a notice is illegitimate.

In a typical sentiment, one OSP described it as "betting the company" every time they decide to leave material up because a notice is false.

Given the risk of liability and the resource-intensive nature of substantive review, what is the rationale for continued investment in the process? Why, in particular, should OSPs spend inordinate time and energy on a small number of edge cases? The answer overwhelmingly given by OSP legal staff—particularly those running UGC sites—was that they feel obliged to combat abuse of the notice system, which can damage not only the expressive rights of individual users but also the larger user environment that sustains the OSP. There is a commercial logic to these choices. Several respondents observed that their companies depend on community good will, which, if lost, would push members toward other substitute services. But a deeper rationale was, for lack of a better word, cultural. OSPs described the enabling of transformative use, re-use, and creative appropriation of cultural materials as deeply intertwined with expressive rights. The majority expressed commitments to protecting the DMCA Classic concept of procedural balance between users and rightsholders, of which the notice and takedown process is the flawed but also last best representative.

[121] A handful of cases have tested the scope of § 512(f). *See, e.g.*, Online Policy Grp. v. Diebold, Inc., 337 F. Supp. 2d 1195, 1205 (N.D. Cal. 2004) (holding that Diebold materially misrepresented its claim in violation of § 512(f), suggesting it sought to use the DMCA "as a sword to suppress publication of embarrassing content rather than as a shield to protect its intellectual property"). Challenges regarding the meaning of "good faith" in sending notices, in contrast, have generally favored the rightsholders. *See* Rossi v. Motion Picture Ass'n of Am., Inc., 391 F.3d 1000, 1003-4 (9th Cir. 2004) (holding that the "good faith belief" requirement is a subjective standard, and a full investigation into the accuracy of the claim is not required); Tuteur v. Crosley-Corcoran, 961 F. Supp. 2d 333, 343-44 (Mass. 2013) (holding that "in enacting the DMCA, Congress did not require that a notice-giver verify that he or she had explored an alleged infringer's possible affirmative defenses prior to acting, only that she affirm a good faith belief that the copyrighted material is being used without her or her agent's permission."); *but see* Lenz v. Universal Music Corp., 2016 U.S. App. LEXIS 5025, at *16 (9th Cir. Mar. 17 2016) (holding that fair use is "authorized by law" and a copyright holder must consider the existence of fair use before sending a takedown notification under § 512(c)).

[122] In addition, counter notices are vanishingly rare in any case; further OSPs routinely limit liability from their users via terms of service. *See infra* Section III.C.4.

4. Counter Notices: Inadequate and Infrequently Used

By its structure, section 512 mostly leaves due process for targets to the privately adjudicated notice-and-takedown process.[123] The main mechanism is the DMCA's "counter notice."[124] Where they apply, the counter notice provisions require OSPs to give targets notice of the complaint against them; targets can respond with a counter notice challenging the takedown.[125] If the target submits a counter notice, the OSP forwards it to the rightsholder.[126] The rightsholder then has ten days to decide whether to sue the user. If a suit is filed, the content stays down pending the outcome.[127] If no action is taken within the ten days, the OSP may restore the content and retain safe harbor protection.[128] While some rightsholders expressed some faith in the counter notice process, OSPs mostly considered it a dead letter—impractical and rarely used. All OSPs and at least one rightsholder agreed that the counter notice procedure's practical ability to protect targets is limited.[129] All agreed that the process has major deficiencies.

First, the counter notice provisions are limited in both structure and practice. The obligation to pass on a complaint does not apply to connectivity providers or to search providers. Further, OSPs that are required to pass on notices typically do not wait for a potential counter notice before takedown.[130] Instead, the material is typically removed, and then may later be restored in response to a counter notice.

Second, by all accounts, the actual use of counter notices is extremely infrequent. Only one respondent among both service providers and rightsholders reported receiving more than a handful per year. Many—including some large services handling thousands of notices per year—reported receiving none.

While OSPs typically inform their users about the procedures for sending counter notices, many do so with considerable ambivalence. Several observed that the typical target of a DMCA complaint has "little or no knowledge of copyright law," and little capacity to make informed estimates of the risks attendant on filing a counter notice, including the risk of

---

[123] *See* 17 U.S.C. § 512(g) (2012). In addition to the counter notice provisions of § 512(g), targets also have recourse to the provisions of § 512(f) which provide liability for damages, including costs and attorney fees, incurred as a result of the service provider relying upon such misrepresentation in removing or disabling access to material. § 512(f). However, as described in Section III.C.3, these provisions have thus far proved weak and unlikely to result in significant recovery.

[124] 17 U.S.C. § 512(g).

[125] Under the statute, OSPs must make a counter notice process available to "subscribers"—a term that narrows the application to § 512(c) services such as UGC sites and storage providers, but not, for example, search engines. *Id.*

[126] *Id.*

[127] *Id.*

[128] Under section 512, an OSP may be liable to a subscriber for disabling access to or removing material if it does not notify the subscriber that the material has been taken down and, upon receipt of a counter notice, replace the material in not less than 10 days nor more than 14 days following the receipt of the counter notice (providing the notice sender has not filed an action seeking a court order to restrain the subscriber from the allegedly infringing activity). *Id.* However, many OSPs write terms of service specifying that they do not need to reinstate material that has been removed.

[129] As noted above in Section III.B.2.a, however, rightsholders also pointed to the counter notice procedure as providing protection to targets from mistaken or abusive notices.

[130] *Id.*

One described counter notices as "irrelevant in [nearly all cases]" because the language in a typical notice is "really threatening...[Users] are too afraid."

copyright liability and the risk of liability for misrepresentation under section 512(f). One described counter notices as "irrelevant in [nearly all cases]" because the language in a typical notice is "really threatening...[Users] are too afraid."

Several OSPs expressed concern about doing anything that might encourage users to assert their rights—even when a notice is clearly invalid—because of the power imbalance between senders and users. In one OSP's view, the prospect of sending users up against media company attorneys backed by statutory copyright penalties "eviscerated the whole idea of counter notice." One rightsholder respondent agreed that "there is a real imbalance of power in the event of a lawsuit. You don't want to go up against [a major entertainment company's] lawyers." Another OSP respondent portrayed the counter notice procedure as a threat to user privacy because it requires the user to disclose both name and address to the sender.[131]

Timing is a third problem cited by both OSPs and rightsholders. As rightsholders press for faster takedown from the major services, some DMCA Auto OSPs now respond to most takedown requests in minutes.[132] Action on counter notices, on the other hand, is still measured in days or weeks. Service providers must restore content based on a valid counter notice no later than fourteen days after receipt, but also, *no sooner than ten days*—the period in which the rightsholder must decide whether or not to sue.[133] OSPs expressed concern about this statutorily mandated delay, pointing out the potentially dangerous effects on expression or competition. As one OSP described it, ten to fourteen days represents "an eternity on the Internet" for small businesses, for community sites where content has a short lifespan, or for political speech (as the McCain presidential campaign learned when a number of its commercials were pulled from YouTube in October 2008[134]).

Some OSPs were more concerned about liability for putting back material than they were about leaving it down. They perceived a risk in reinstating material based on a counter notice; not all were fully comforted that they would be able to maintain the safe harbor if the material were challenged after 10 days. One provider argued that the statutory language is not sufficiently clear that OSPs other than hosting providers are protected for putback at all.[135] Another summed up the DMCA's relatively broad latitude for notices targeting non-infringing content and the limited nature of the counter notice process as making the DMCA "an excellent mechanism for censorship."

---

[131] One respondent described "cases where allegedly abusive ex-husbands have filed DMCA complaints against images their ex-wife had posted as a means of attempting to get her current address...". This OSP suggested setting up a "proxy service which allows someone to contest the charges and accept legal responsibility if found guilty without having to reveal their identity to the filing party."

[132] Nevertheless, rightsholders cited delays in takedown as an issue in their efforts to contain and manage rapid-fire proliferation of content online: "once it is out it is out."

[133] 17 U.S.C. § 512(g).

[134] Ctr. for Democracy & Tech., Campaign Takedown Troubles: How Meritless Copyright Claims Threaten Online Political Speech (2010), https://cdt.org/insight/campaign-takedown-troubles-how-meritless-copyright-claims-threaten-online-political-speech/.

[135] This concern is grounded in the "putback" language of section 512(g), which applies to "material residing at the direction of a subscriber." 17 U.S.C. § 512(g)(2). For OSPs, the fact that a court has never addressed this issue is beside the point: they do not want to go to court at all.

On the other side of the coin, rightsholder groups and at least one OSP also described the counter notice process as subject to abuse by pirates—especially those operating from other jurisdictions. More than one respondent described bogus counter notices from obvious foreign copyright pirates claiming the right to post infringing material. For example:

> "We have only received only seven [counter] notices in the last two years (we have sent nearly 9,000 notices to Google). Two were a result of administrative errors on our end. Five were from Russian or Ukrainian torrent sites that knew that there was no chance that we would sue them in their jurisdiction."

Several rightsholders observed that determined infringers are not deterred by the threat of 512(f) penalties for misrepresentation, particularly if they feel immune from U.S. lawsuits in the safety of another jurisdiction. They described these targets as exploiting the fact that offshore infringement is hard to reach with lawsuits. Indeed, bogus Ukrainian counter notices appear to have a special place in the tiny counter notice universe. Regarding Google's counter notice procedures for YouTube takedowns, another rights enforcement agent observed:

> "Suppose a Ukrainian kid uploads one of our newly released movies to YouTube. It gets flagged by Content ID. We evaluate and send a takedown request. YouTube notifies the kid of the takedown and the kid responds with a counter notice claiming that he holds worldwide distribution rights for the movie. Now we have 10 days to decide whether to sue the kid in Ukrainian court. If we don't, the movie goes back up. This has happened a couple times."

In tension with what OSP respondents reported about their fears of liability for putback, one rightsholder stated that most OSPs simply reinstate the content when presented with a counter notice and decline to make a judgment call about the merit of the claim in the counter notice. Overall, the counter notice practice was described as less than satisfactory by nearly all respondents.

5. Repeat Infringer Policies and "Strikes"

Several OSPs noted the practical interplay between counter notice provisions and section 512's requirement for repeat infringer policies, which mandates OSPs to adopt policies for terminating the accounts of "repeat infringers."[136] This is one of section 512's more controversial provisions. It applies to all types of OSPs covered by section 512, and the remedy—termination of an online account or even basic Internet connectivity service—could potentially deal a severe blow to expression in the online world, where private intermediaries provide the main platforms for speech. Despite this, however, the definitions of both "repeat infringer" and "appropriate circumstances" for termination are unclear in the statute, as are the scope or duration of the penalty. Both the controversy and lack of clarity are reflected in practice, where OSPs' implementations vary.

---

[136] Section 512(i)(1)(A) requires the "reasonable" implementation of a "policy that provides for the termination in appropriate circumstances of… repeat infringers." 17 U.S.C. § 512(i)(1)(A). Courts have regularly applied a three-prong test enunciated by the Ninth Circuit in *Ellison v. Robertson* to determine whether service providers meet the safe harbor eligibility requirements of section 512(i)(1)(A). *See* Ellison v. Robertson, 357 F.3d 1072 (9th Cir. 2004). Under the test, service providers must (1) adopt a policy that provides for the termination of service access for repeat infringers in appropriate circumstances, (2) inform users of the repeat infringer policy, and (3) implement the policy in a reasonable manner. *Id.*

While the statute does not provide much guidance about how to comply, most OSPs have adopted models that assign users "strikes" each time they are the objects of notices; too many strikes, and the account is terminated. Typically, a valid counter notice will remove a strike. One OSP that permanently terminates users after three strikes noted that although users typically forgo the opportunity to file a counter notice for their first two notices, when users receive a third strike they become much more animated about sending counter notices in order to avoid termination.

> ***One provider indicated that, after experimenting with different repeat-infringer policies, he had concluded that the only safe interpretation was to apply a literal reading of the word "repeat" by suspending any user who is targeted by a second notice.***

Beyond this basic "strike" approach, it is hard to discern a particular trend in policies. Some are very conservative. At this end, one provider indicated that, after experimenting with different repeat-infringer policies, he had concluded that the only safe interpretation was a literal reading of the word "repeat." His service suspends posting rights for users who receive a second notice. OSPs more commonly expressed discomfort with such categorical actions. Notices, after all, remain accusations, not proof of infringement. Most OSPs require more before terminating an account. A few allow users to get out of the "penalty box" of termination with good behavior; others do not.

Some OSPs' strike policies are quite simple; others can be very complicated and unique to the OSP. This makes it too identifiable for us to go into much detail based on our interviews, but, some publicly available examples provide an idea of the type of complexity of some policies. Pinterest, for example, distinguishes notices from strikes, and gives senders the option of either checking a box to indicate that a sender is "asking Pinterest to assign a strike against the user" or sending the takedown notice without it counting as a strike.[137] Google treats DMCA notices sent to YouTube as strikes against users, but does not count Content ID matches as strikes.[138] It distinguishes between copyright and "community standards" strikes, which can also result in termination.[139] Google's set of user remedies is similarly complicated; it ranges from the contestation of notice-based strikes through formal counter notice, to the retraction of strikes via appeals to the rightsholder, to the expiration of strikes if the user completes online "copyright school" and receives no further strikes in a six-month period.[140] Three strikes lead to the permanent termination of the user's account, with all uploaded files removed.[141] Most systems our respondents described are not as complicated as that; however

---

137 *See Copyright Infringement Notification*, PINTEREST, *supra* note 117. Pinterest's web form allows notice submitters to check a box indicating that the sender is "asking Pinterest to assign a strike against the user" who posted the content that the sender is requesting be removed. *Id.*

138 YouTube users receive strikes when a video is taken down because a copyright owner sent Google a "complete legal request" to remove the content. *Copyright Strikes Basics*, YOUTUBE, https://support.google.com/youtube/answer/2814000 (last visited Feb. 5, 2016). Three strikes lead to the permanent termination of the user's account, with all uploaded files removed. *Id.* While Content ID matches do not count as strikes, invalid disputes of Content ID blocks can result in copyright strikes. *Keep Your YouTube Account in Good Standing*, YOUTUBE, https://support.google.com/youtube/answer/2797387 (last visited Feb. 5, 2016).

139 *Community Guidelines Strikes*, YOUTUBE, https://support.google.com/youtube/answer/2802032?hl=en&vid=1-635763131740164071-4037538024 (last visited Feb. 5, 2016).

140 *Copyright Strikes Basics*, YOUTUBE, *supra* note 138.

141 *Id.*

it is clearly the case that, for users engaged with a wide range of services, the contours of rights and due process around copyright issues have become very complex.

In a further complication, "repeat infringer" requirements apply to all the different services covered under 512(a) and 512(c)—Internet connectivity providers and hosting or storage services, respectively. However, early jurisprudence on 512(a) strongly affirmed general safe harbor protection when the ISP acts solely as a data conduit; responding to notices is not required.[142] This mirrors traditional conceptions of "mere conduits" as neutral third parties not liable for the bad acts of those who use them, but complicates the relationship between notice sending and repeat infringer policies, which often use notices as a trigger.[143] For some time, ISPs received large numbers of 512(a) "takedown notices" from major rightsholders targeting peer-to-peer file sharers. Though section 512 did not oblige ISPs to respond to these notices, rightsholders argued that the notices provided evidence of repeat infringing activity and that ISPs should terminate users' accounts based on them. Indeed, automated mapping of user IP addresses across peer-to-peer networks was the first major use case for automated notices. In interviews, ISPs reported receiving upward of 1 million notices per year during this period—nearly all peer-to-peer related.

This argument waxed and waned for some time, but as further discussed below, the major rightsholder groups largely abandoned the DMCA as a pressure point against ISPs in favor of legislative strategies and quasi-private deal making in the form of the Copyright Alert System.[144] But as notice costs have fallen, other enforcement organizations have moved in to fill the niche. For example, one ISP reports that it continues to receive around 12 million notices per year for its connectivity services from REOs like Rightscorp, which asks ISPs to forward private settlement offers to alleged infringers.[145] Because the ISP views these notices as invalid, they are systematically deleted. This can still come at some cost. Indeed, the respondent that reported 12 million notices a year described *manually* deleting thousands of notices per day after individually verifying that none of them referred to 512(c) services, for which the ISP could be liable.

One ISP described manually deleting thousands of notices per day after individually verifying that none of them referred to 512(c) services.

[142] In *RIAA v. Verizon*, the DC Circuit Court of Appeals affirmed that ISPs were not subject to notice and takedown requirements due to the impossibility of locating and taking down materials in transit across ISP networks. Recording Indus. Ass'n of Am., Inc. v. Verizon Internet Servs., Inc., 351 F.3d 1229 (D.C. Cir. 2003). Because § 512(a) notices were invalid, moreover, rightsholders could not avail themselves of injunctions available under § 512(j) to force ISPs to reveal the names of accused customers. *Id.* Subsequent RIAA efforts to compel bulk disclosure of customer names in the context of lawsuits against file sharers also failed. For a comprehensive history, see Elec. Frontier Found., RIAA v. The People: Four Years Later, https://w2.eff.org/IP/P2P/riaa_at_four.pdf.

[143] This idea is not without controversy, as notices are accusations rather than proof of infringement. For a discussion of what constitutes an "infringer" under 512(i), *see* David Nimmer, *Repeat Infringers*, 22 J. of Copyright soc'y U.S.A., 167 (2004).

[144] *Infra* Section III.D.1.c.iv.

[145] For example, Rightscorp sends notices to ISPs to forward to their customers that offer to settle cases of alleged copyright infringement through peer-to-peer networks for a small sum in lieu of pursuing litigation. *See Pay a Notice*, Rightscorp, http://www.rightscorp.com/notices/pay-a-notice (last visited Feb. 5, 2016).

6. Transparency Reporting and Public Archiving of Notices

Despite the commonalities in OSPs' experiences, they uniformly reported having little knowledge of other service providers' notice and takedown practices. Knowledge about how services manage notice and takedown across the Internet sector remains remarkably limited. Fourteen years after the founding of the Chilling Effects Clearinghouse (now Lumen), the public archive still has only two major contributors: Google, which began archiving notices to Web Search in 2002 after controversy surrounding takedown requests from the Church of Scientology,[146] and Twitter, which began archiving in 2012.[147] Google has since added services beyond Web Search to its Lumen submissions,[148] but notably does not include YouTube, which is squarely within the litigation-intensive audiovisual services sector and which has been the subject of hard-fought litigation. Beyond public archiving, Google also makes notice metadata for its Search service searchable via its Transparency Report, which began archiving in 2011.[149]

One might justifiably have thought that Google's transparency efforts would create a wider norm in the field. Certainly the problem of questionable notices has not noticeably changed. Takedown requests targeting non-infringing content consistently came high on the list of concerns among the OSPs we interviewed. But notice archiving and transparency reporting have yet to become norms. Google's decision to publicly archive its takedown requests and related data was likely made easier by the weak and unsympathetic case of the Scientologists, the limited volume of takedown requests it received at the time, and the fact that section 512 was not yet a heavily litigated area. This combination of incentives proved to be an exception rather than the rule.[150]

While transparency has advocates among some OSPs and rightsholder groups, public reporting has, for most, remained a subject of discussion rather than action. Although a number of companies recently began to publish transparency reports in the wake of the Snowden revelations (including Microsoft, Facebook, LinkedIn, Yahoo!, and others), most are limited so far to documenting the frequency of *government* requests for user information rather than copyright complaints. Only a handful provide data on

While transparency has advocates among some OSPs and rightsholder groups, public reporting of notices and responses to them has, for most, remained a perpetual subject of discussion rather than action.

[146] Church of Scientology. Gallagher, *supra* note 11. The takedown requests were widely viewed as an effort by the Church of Scientology to silence critics. Google initially removed and later restored the offending search results. Regarding the controversy, see *Operation Clambake*, WIKIPEDIA, http://en.wikipedia.org/wiki/Operation_Clambake (last updated Nov. 7, 2015). For the relevant notice to Google, see *Google Asked to Delist Scientology Critics (#1)*, CHILLING EFFECTS (Mar. 8, 2012), http://www.chillingeffects.org/notice.cgi?NoticeID=232.

[147] Other organizational submitters include Proxy.sh, Stack Exchange, Wikimedia Foundation, WordPress, Medium, Kickstarter, Stripe, and Tucows. E-mail from Adam Holland, Project Coordinator, Lumen (July 29, 2014, 10:10 PST) (on file with authors).

[148] Google contributes notices for the following Google products to Lumen: App Engine, Blogger, Chrome Web Store/Extensions Gallery, Code, Currents, Drive and Docs, G+, Geo – 3D Warehouse, Geo – Panoramio, Google Cloud Storage, Google+ Local, Google Profiles, Groups, Image Search, Orkut, Page Speed Services, Picasa, Sites, and Web Search. E-mail from Shantal Rands Poovala, Google Inc. (May 9, 2014, 11:43 PST) (on file with authors).

[149] *Transparency Report*, GOOGLE, *supra* note 12.

[150] Additionally, Google's financial and engineering resources make it an outlier among OSPs.

copyright takedown. At the end of 2015, this group included Mega, Microsoft, Reddit, Twitter, Wikimedia, and WordPress.[151]

Several OSPs told us that this lack of transparency leaves them in the dark about how others manage the DMCA's various ambiguities, at times leading them to make decisions and set policies conservatively. In general, OSPs agreed that more information would support good internal practices and potentially improve public relations by anchoring commitments to transparency and user rights.[152] Yet in most cases these benefits were not enough to overcome hesitation: public reporting remains rare.

Resource constraints were the most widely cited factor in decisions not to archive notices or publish transparency reports. Despite the general sense that transparency reporting would please some users, some OSPs saw little or no actual user demand for transparency reporting, making it difficult to justify the resources required. Small-to-medium sized OSPs, especially, described their notice and takedown procedures as geared toward minimizing the staffing and infrastructure costs required to protect the OSP, its users, and rightsholders, and substantial cost increases as untenable. Several also described the challenges inherent in converting legacy systems for handling notices into systems that could handle transparency reporting as major obstacles to change. Because relatively few small-to-medium-sized services receive large numbers of notices—or, in a few cases, have only recently begun to receive large numbers—few have back-end systems designed to manage complex notice workflows or generate regular analyses. Few have systems that can unify and manage the flow of emails, PDFs, Word documents, faxes, and letters that still make up a portion of notices for many providers—including those that have form-based submission options. This is true even of some very large OSPs, which have greater resources but also greater complexity in their services and infrastructure.

OSPs that host UGC generally had additional reasons for reluctance. The direct hosting relationships these services have with the targets of takedown requests lowered their comfort with transparency reporting. Some hesitated to expose users whose content has been alleged—but not proved—by the sender to be infringing. Search engines like Google Web Search, on the other hand, have no client relationships with users and are thus less encumbered by user privacy or related user considerations. Search providers thus considered publicly archiving notices as comparatively unproblematic but sometimes still hesitated in light of the potential controversy of identifying senders.[153]

Several OSPs also worried that transparency reporting and public archiving could trigger negative attention from rights enforcement groups, exposing them to high-volume sending or even litigation. Google, in this respect, was described as a cautionary tale.

Several respondents expressed concern that transparency reporting and public archiving could trigger negative attention from rights enforcement groups.

[151] *Supra* note 12.

[152] For an example of this type of messaging, see *Transparency is Important to Us, and Today, We Take Another Step Forward*, Reddit (May 13, 2015), https://www.reddit.com/r/announcements/duplicates/35uyil/transparency_is_important_to_us_and_today_we_take/.

[153] *Statement from: Copyright Alliance CEO Sandra Aistars, to Committee on the Judiciary Subcommittee on Courts, Intellectual Property and the Internet U.S. House of Representatives RE: "Section 512 of Title 17*, Copyright Alliance (Mar. 13, 2014), https://copyrightalliance.org/2014/03/statement_copyright_alliance_ceo_sandra_aistars_committee_judiciary_subcommittee_courts#.VeztJJ3BzGc [hereinafter *Statement from: Copyright Alliance CEO Sandra Aistars*].

These OSPs appreciated the goals behind the Transparency Report, but worried that it both contributed to Google's status as a magnet for automated notices and that it invites critics to complain that there is too much infringement on Google Web Search.[154] One OSP indicated that it does not issue a transparency report due to, in part, "the increased potential of being subject to a lawsuit for doing so." Unsure about why rightsholders send large numbers of notices to some sites and not others, some DMCA Classic OSPs prefer to remain "under the radar." The potential costs of attracting large numbers of notices of unknown quality, and the fact that it would prohibit substantive notice review, were primary concerns for these OSPs. Smaller OSPs operating on lean budgets worried about potentially large cost increases if transparency reporting were to attract automated notices. Larger services that received relatively low numbers of notices were especially concerned that even documenting the apparent gap between their experience and that of other large services could attract the attention of opportunistic REOs or litigants. Many worried that such attention would force them to either dramatically increase their spending on compliance or jettison the human review process. Some OSPs expressed the concern that transparency reporting would send an unintended negative political message. One OSP even worried that archiving with Lumen could be viewed as a provocation, indicating that its service had "no strong desire to thumb our noses at rightsholders/creators with legitimate interests and concerns."

One OSP even worried that archiving with Lumen could be viewed as a provocation, indicating that its service had "no strong desire to thumb our noses at rightsholders/creators with legitimate interests and concerns."

Concerns that transparency might invite negative attention are not entirely without basis. Lumen has come under growing pressure from some rightsholders precisely because of its status as a public archive.[155] Sandra Aistars, then representing the Copyright Alliance, expressed this viewpoint in her 2014 Congressional testimony, arguing that the Lumen site (then Chilling Effects) "unfairly maligns artists and creators using the legal process created by section 512 as proponents of censorship" and as a separate concern, "has effectively become the largest repository of URLs hosting infringing content on the internet."[156]

On the other hand, several rightsholder respondents also expressed interest in stronger transparency practices as a means of demonstrating the fairness and accuracy of their enforcement practices, though they are reluctant to reveal details about enforcement methods that might enable counter-strategies by file-sharing sites. This interest included,

[154] At least one commentator has made this criticism. *See, e.g.*, BOYDEN, *supra* note 6.

[155] See ROBERT LEVINE, FREE RIDE: HOW DIGITAL PARASITES ARE DESTROYING THE CULTURE BUSINESS, AND HOW THE CULTURE BUSINESS CAN FIGHT BACK (2012) 86 (stating that the Chilling Effects site "presents lawful requests from creators to stop unauthorized distribution of their works as a threat to free speech."). In keeping with the accusation that Lumen abets infringement, some rightsholders have even sent DMCA notices to Lumen, demanding that it remove DMCA takedown notices. Ellen Seidler, *I Sent Chilling Effects a DMCA Takedown Notice*, VOX INDIE, (May 7, 2015), http://voxindie.org/i-sent-chilling-effects-a-dmca-takedown-notice/.

[156] *Statement from: Copyright Alliance CEO Sandra Aistars*, *supra* note 153. In an attempt to balance the importance of making takedown request information available to study, research, and journalism, while still addressing the concerns of people whose information appears in the database, Lumen has recently removed notice pages from search engine results. Ernesto, *Chilling Effects DMCA Archive Censors Itself*, TORRENTFREAK (Jan. 10, 2015), https://torrentfreak.com/chilling-effects-dmca-archive-censors-itself-150110/.

in the case of one large rightsholder, the possibility of releasing its own transparency report. Nor do anti-transparency positions appear to be widespread in the policy community, where the USPTO and other official bodies have repeatedly called for more empirical research on takedown.[157]

Despite these attitudes, in an environment with weak incentives for transparency, uncertain back-end costs, and a potential for greater exposure to rightsholder campaigns, the practice of notice and takedown has—with a few exceptions—largely remained a black box. Although the shift toward form-based submission of notices is likely to provide a better technical basis for reporting in the future, the incentive structure for reporting remains unclear for many of our respondents. The Google precedent from the early days—marshaling transparency as a defense against abusive notices and user criticism of takedown—was created in a different political and legal environment. Today, regular litigation over section 512's requirements and obligations reinforces secrecy and ambiguity regarding practices on all sides. Some OSPs fear any step that could upset the fragile equilibrium in which they operate. Desire for transparency has not been enough to tilt the balance for most providers.

In an environment with weak incentives for transparency, uncertain back-end costs, and a potential for greater exposure to rightsholder campaigns, notice and takedown practice overall has largely remained a black box.

D. DIVERGENCE IN TAKEDOWN PRACTICE: MOVING TO DMCA AUTO, DMCA PLUS, AND BEYOND

In recent years, notice and takedown practice has diverged. On one side of this split, the DMCA Classic OSPs process manageable numbers of takedown requests, generally using human review. On the other side, rightsholders use automated systems to send, and OSPs process, notices on a massive scale.

The challenge of managing takedown on much larger scales has required changes in OSP practices. Some of these practices—the group we call "DMCA Auto"—retain the basic notice and takedown process, but implement it on a large scale. Other practices—the group we call "DMCA Plus"—include attempts to proactively prevent infringing material from making its way onto (or staying on) an OSP's system. The distinction is legally relevant: DMCA Auto OSPs are still following their obligations under section 512 to accept takedown requests and remove specific instances of identified infringement, however massive the scale. DMCA Plus practices, however, move beyond the statutory requirements and supplant notice and takedown as the primary mechanism for managing copyright disputes. In the DMCA Plus realm, decisions about content are made preemptively, regardless of "red flag knowledge."[158]

There is a considerable overlap between DMCA Auto OSPs and DMCA Plus OSPs. Most DMCA Plus OSPs we interviewed also employ DMCA Auto measures. Numerically, the DMCA Auto and DMCA Plus groups together represent a minority of our respondents—only nine of twenty-nine employed any of the enforcement measures described in this

---

[157] Seng's 2014 study, for example, relied on data crawled from the Lumen (then Chilling Effects) site. Seng, *supra* note 5, at 378-83.

[158] We are grateful to Bill Rosenblatt, who suggested dividing automated practices into "reactive" practices, which follow the statute and involve OSPs responding to notices from rightsholders, and "proactive" practices, that move beyond the statute and filtering and other *ex ante* infringement policing.

section (sites that do not employ such measures include some with heavily trafficked sites as well as sites with large staff). The nine, however, includes some of the dominant Internet services in their respective areas. While the small number of OSPs that fall into either category precludes our ability to discern a clear trend, it may be that DMCA Auto OSPs tend to collapse into DMCA Plus measures as they attempt to manage very large numbers of takedown requests.

Among the relatively few OSPs that have implemented these measures, even fewer are willing to discuss their notice handling practices in detail. In many cases, even the broader outlines of these policies are not publicized and do not circulate widely within the OSP community. Because the adoption of new measures is almost always the result of pressure from rightsholder groups, there is a premium on the discreet resolution of disputes. However, the general contours of these systems were relatively clear.

1. From DMCA Auto to DMCA Plus Enforcement Measures

Although automated notices have failed to prevent infringing content from appearing online, they have pushed OSPs that receive them to turn to DMCA Auto and DMCA Plus measures.

DMCA Auto and Plus practices are generally—though not always—born from the pressure to manage large numbers of automated takedown requests and/or threats of litigation. Although automated notices, by general agreement, have failed to prevent infringing content from appearing online, they have pushed the OSPs that receive them to turn to DMCA Auto and Plus measures in an effort to reach a new detente with rightsholder groups and to assert control over the copyright disputes on their services.[159]

A few other OSPs also expressed a sense of pressure to adopt some of the DMCA Plus practices, regardless of whether they were currently targeted by large-scale notice sending. Many OSPs viewed major changes in the scale of the notice and takedown process as a break with the balance of rights, responsibilities, and remedies the process was designed to maintain. Some aspects of these techniques, all agreed, can supplant the DMCA's procedural remedies and the accompanying protections for users. Some OSPs worried that wide adoption of DMCA Plus steps beyond the statutory requirements could undermine the safe harbor by shifting norms around reasonable or standard measures and extend enforcement beyond its statutory constraints. OSPs expressed concern both that unaffordable *de facto* standards could develop, and that preemptive DMCA Plus measures challenged the "publish first, enforce after" rubric of the DMCA.

Pressure from increased numbers of notices and threats of litigation appears to be focused on OSPs in sectors where copyright issues are highly contested, such as search, cloud storage, video, and music services. OSPs in the crossfire of these overarching disputes were more likely to introduce DMCA Auto measures and in some cases, to move on to DMCA Plus measures that are not required for the safe harbor but give rightsholders broader enforcement power.

[159] For an excellent survey of DMCA Plus measures and Para DMCA measures, *see* Annemarie Bridy, *Copyright's Digital Deputies: DMCA-Plus Enforcement by Internet Intermediaries*, RESEARCH HANDBOOK ON ELECTRONIC COMMERCE LAW (John A. Rothchild ed., forthcoming 2016).

a. DMCA Auto Practice

DMCA Auto practice does not substantively depart from the traditional notice and takedown regime—it merely automates the process in order to manage floods of requests. DMCA Auto OSPs described, for example, developing automated workflows to match claims to content, remove or block relevant material, triage edge cases for human review, and—for 512(c) services—notify users.

The practical features of DMCA Auto measures, however, are very different from those of human-driven DMCA Classic measures. The vast majority of infringement claims DMCA Auto and DMCA Plus OSPs receive are not substantively reviewed—either by the senders, who rely largely on title matches and similar proxies to identify copyrighted material, or by the recipient OSPs, which can, at best, triage small percentages of notices for human review. Though substantive review of claims is limited, the notices still generally trigger OSP knowledge. Unable to evaluate every takedown request or fail to act on valid ones without risking their safe harbor protection, OSPs may take down material even where there is doubt about the substance of the claim. As one observed: "At any real volume, the [notice and takedown] process, as written, is impossible."

One OSP observed: "At any real volume, the [notice and takedown] process, as written, is impossible."

Many rightsholders and some OSPs viewed these systems primarily through the lens of efficiency—of facilitating and ensuring compliance with the DMCA process on its new expanded scale. For some OSPs, however, compliance at this level has not produced a new equilibrium, but further pressure from rightsholders to implement DMCA Plus measures.

b. Transitional Practices: Trusted Sender Programs and Direct Takedown Privileges

In addition to developing automated systems to process large volumes of incoming notices, some DMCA Auto OSPs have responded to them by developing "trusted user" programs that facilitate bulk notice sending for "trusted" senders and fast-track takedown for these senders. These programs vary by OSP and, depending on their design, may straddle the line between DMCA Auto measures used to accomplish section 512's requirements efficiently and DMCA Plus measures that go beyond the requirements for safe harbor protection. On the DMCA Auto side, OSPs create special processes for "trusted senders" in notice workflows that simplify bulk submission, sometimes with little or no substantive review—a practice that is within the statutory framework but accomplished on a massive scale. On the DMCA Plus side, OSPs have gone beyond the statute's requirements by providing senders with backend access and accompanying takedown privileges.

"Trusted" sender programs were the most common such measure described to us. Though they vary in detail, all such programs facilitate bulk notice sending and streamline the removal process, often with limited or no OSP review.[160] One OSP

One OSP described a process that is entirely automated: When a "trusted" sender submits a takedown request, the automated system immediately removes the content and notifies the targeted user.

[160] In general terms, "trusted" programs are the mirror image of policies to subject individual or one-off senders—whose notices OSPs consider more likely to be suspect—to extra scrutiny. *See supra* Section III.D.1.b.

described an entirely automated process: When a "trusted" sender submits a takedown request, the automated system immediately removes the content and notifies the targeted user. Others conduct limited review, typically using automated processing to flag potentially problematic requests (such as requests to remove home page URLs), which may then trigger human review.

In other cases, OSPs provide senders with access to backend systems that go beyond streamlining the notice-handling process, and into DMCA Plus territory, by allowing senders to remove content directly. Some sites allow "trusted" senders to remove content directly from their hosting services without formal notices, identification of the infringed work, user notifications, or review. According to respondents, these direct take down arrangements between DMCA Plus OSPs and senders generally arise in response to a sudden increase in the volume of notices received by a service or threats of litigation. We heard from several respondents that these types of backdoor access agreements are relatively common among digital music file-hosting services.[161] For example, one REO described having direct take down privileges with eighteen file-hosting services.

Some DMCA Plus OSPs provide senders with access to backend systems that allow senders to remove content directly.

c. DMCA Plus Practices

A few OSPs had moved from DMCA Auto practice squarely into the world of DMCA Plus practices. Most commonly, these included site-wide removal based on hashing technology or *ex ante* filtering to keep certain files off an OSP's platform altogether. Much more rarely, OSPs implemented "staydown" measures intended prevent material that had been removed in response to a takedown request from being reposted another time or by another user. Finally, some OSPs and rightsholders have entered into side agreements that supplement or replace the section 512 process.

With the exception of side agreements, which are too diverse to categorize neatly, these practices both go beyond statutory requirements and reverse the usual burdens. The DMCA does not require OSPs to monitor the activity of their users for infringement,[162] or to take proactive action against users before they gain "red flag knowledge" of infringement. As with copyright more generally, the initial burden of identifying infringements and sending notices belongs to copyright holders. Section 512 extended this to online disputes, in part, to avoid creating incentives for OSPs to prospectively police user expression. Unsurprisingly, many OSPs expressed strong reservations about implementing DMCA Plus measures, some of which look like prospective policing, and most of which shift cost burdens toward OSPs.

Like DMCA Auto measures, rightsholders largely viewed DMCA Plus measures in terms of efficiency. Rightsholders' motivations for pressing OSPs to adopt DMCA Plus measures were clear. Though all agreed that notice and takedown remains central to online copyright enforcement, none viewed it as adequate for addressing large-scale online infringement. Among the OSPs, motivations were more complex. For some, steps beyond the DMCA were part of a process of experimenting with better ways of balancing rightsholder and user needs. For some, they were concessions to rightsholder pressure, adopted to deflect threats

[161] *See, e.g.*, Ernesto, *Universal Music Can Delete Any SoundCloud Track Without Oversight*, TORRENTFREAK (July 3, 2014), http://torrentfreak.com/record-labels-can-remove-soundcoud-tracks-without-oversight-140703/.

[162] 17 U.S.C. § 512(m) (2012).

of litigation. For some, they were ways of re-asserting some control over copyright disputes on their services, which the blanket takedown response to automated sending did not afford. Often motivations were a combination of the three.

i. Hash-Matching and Site-Wide Removal

Section 512 specifies only that an OSP must "remove, or disable access to, the material" in response to a proper notification. The latter phrase has been conventionally understood to refer to removing individual files or links specified in the notice; courts have generally agreed.[163] While trusted partner programs provide rightsholders with efficient mechanisms for noticing and removing specified materials, some rightsholder respondents argued that this increased efficiency is not sufficient to manage the proliferation of infringing content online. These rightsholders argued that in order for the notice and takedown system to be effective in addition to efficient, a more expansive definition of what content should be removed in response to requests is required. In this situation, several rightsholder respondents argued, effective takedown requires site-wide removal of the allegedly infringing file. In their view, narrower takedown practice has been rendered ineffective by the rapid repopulation of links and files on file-sharing sites, including rapid community reposting and—in some cases—automated systems for rotating links on linking sites.

Rightsholders' concerns on this issue track a technological shift in cloud storage architectures from individual storage to distributed provisioning. Early services tended to maintain individual copies of files for each user. Naming individual files or links in a notice broadly sufficed to target the modalities of storage and access at most cyberlockers, UGC platforms, and other services through the late 2000s. More efficient cloud architectures, however, often dispense with individualized file storage in favor of maintaining one or a few copies of widely used files, and then apportioning access to as many users as needed. The proper technical approach to removing or disabling access to content in such a context is non-obvious. If an OSP receives notice about a link to an alleged infringing file, should it remove only that link? All links to the file? The file itself? If a targeted file belongs to multiple subscribers, how can one distinguish infringing from non-infringing uses?

OSPs expressed strong reservations about blanket takedown strategies. Several pointed out policy and legal issues that arise with different users of the same file. One user may be making a fair use; another may not. One may have license; others may not. OSPs challenged the idea that they could—or should be able to—tell. For cloud storage and UGC services, user privacy was a major concern.

Still, though most OSPs continue to practice individualized takedown, some have moved toward a middle ground that distinguishes personal storage from features that allow wider sharing. Several described using hash-based matching to remove or prevent all *publicly shared* links to hash-matched files on receipt of a notice for a particular file. In one public example of this approach, when Dropbox receives a DMCA notice, it both disables the identified

[163] Viacom Int'l Inc. v. YouTube, Inc., 718 F. Supp. 2d 514 (S.D.N.Y. 2010), *aff'd in part, vacated in part on other grounds, remanded*, 676 F.3d 19 (2d Cir. 2012); *see also* Wolk v. Kodak Imaging Network, 840 F. Supp. 2d 724, 747 (S.D.N.Y. 2012) (stating that "it would be irresponsible" for the online service provider to assume infringement based on an assumption that one notice of infringement applies to all instances of that content appearing on the website because it may result in blocking others uploading content to which the uploader holds a valid license).

link, and also has an "automated system that prevents other users from sharing the identical material using another Dropbox link. This is all done by comparing file hashes."[164]

In many cases, OSPs described their policies on this point as influenced or dictated by system architecture. Most described architectural challenges to site-wide takedown; legacy designs do not usually accommodate it. One major service among our respondents, however, had a less common legacy site design feature that it now uses to take down "master files" system-wide. This was an early architectural design choice intended to boost server efficiency and not a policy stance—the fact that it facilitated site-wide takedown was merely an accidental secondary feature—but the OSP was able to use it for site-wide takedown.

Rightsholder decisions about whether to request site-wide takedown also exhibit some nuance. Several OSPs that offer senders an option of removing all hash-matched files from the system reported that not all senders elect to use this option. For senders, the decision to remove content may be context-dependent; it is not uncommon for rightsholders to want some uses removed but not others. For example, OSPs have encountered rightsholders engaging in stealth marketing campaigns that involve seeding remixed content, and those that wish to tolerate or support certain fan uses of content. What crosses the line from tolerated to unacceptable use varies from rightsholder to rightsholder, and indeed within rightsholder companies by property. Rightsholder decisions on whether site-wide removal or removal of an individual link or file is appropriate also appear to vary by type of OSP: both rightsholders and OSPs indicated that site-wide removal may be rightsholders' preferred response for file-storage or sharing sites, while they may be more willing to accept individual review on UGC sites.

OSPs expressed uncertainty about whether site-wide takedown was or could become a threshold practice for safe harbor protection. Despite the lack of a legal requirement, some perceived its potential as a new norm as a source of business risk; changing norms could potentially undermine the notion that liability attaches only to specifically identified acts of infringement. Still, several perceived hash matching as 'safer,' from a liability perspective, than traditional takedown. Several view their measures as compromise positions designed to head off stronger rightsholder demands.

ii. Fingerprinting and Filtering

From an enforcement perspective, hash-based systems have a significant limitation: they detect only exact matches to the hashed file and not altered versions of the same material. Slight differences in an image or a song file will produce a different hash value, requiring separate identification and takedown and providing no assurance that other variations will not appear. For example, a hash-based system will fail to detect a match if a song is compressed in a different codec, has a slightly different pitch, is distorted, or has more or less silence at the beginning.

In contrast, fingerprinting-based filtering systems are designed to detect inexact matches, thereby enabling more comprehensive takedown. Filtering systems work by using software that inputs the content file into an algorithm representing it as a set of numbers that represent its perceptual characteristics. These "fingerprints" are computed, registered, and used to

[164] Kyle Orland, *Dropbox Clarifies its policy on Reviewing Shared Files for DMCA Issues*, Ars Technica (Mar. 30, 2014), http://arstechnica.com/tech-policy/2014/03/dropbox-clarifies-its-policy-on-reviewing-shared-files-for-dmca-issues/.

compare against files on an OSP's service, irrespective of variations in the source, codec, quality, or context of use (though handling rules can be put in place for inexact matches).

As with other DMCA Plus measures, OSPs reported that filtering is likely to be adopted under considerable pressure and concern for liability. As one OSP that filters put it:

> "If you are hosting [music or video] content, I don't see how you can deal with that risk without having some sort of content filtering long term. It is not a requirement under the DMCA, but there is too much uncertainty in the DMCA and there is too much risk; it is potentially catastrophic. [Adopting filtering technology] is a reflection of the fact that we don't think that how the DMCA as written and interpreted [offers enough protection from liability]."

On the flip side of these liability fears are substantial concerns about the cost of fingerprinting and filtering technology. Most OSPs do not filter, and all noted that filtering technology is very costly to develop and deploy. Even well-resourced OSPs feared being forced to deploy filtering technologies; many saw them as unattainably expensive. Again, Google was a frequently mentioned example—respondents had heard rumors of Content ID's cost that varied, but all placed it in the tens of millions of dollars, and in some cases much higher.[165]

As an apparent side effect of these high development costs, respondents described the filtering market as dominated by two major proprietary systems: Audible Magic and Content ID. Audible Magic was one of the first commercial fingerprinting and filtering services for audio and enjoyed active promotion by the Recording Industry Association of America ("RIAA") as early as 2004—quickly becoming the *de facto* industry leader. Much of the discussion around filtering focuses on Content ID, Google's internally developed fingerprinting system for YouTube. According to respondents, Google initially signaled that it would license Content ID to others but then backed away from this, leaving Audible Magic to dominate the market for licensed filtering systems. While they vary in important ways, and details are proprietary, Content ID and Audible Magic share some key features. Both allow rightsholders to set rules for how flagged materials are handled, from unequivocal *ex ante* blocking of content to more flexible case-by-case evaluation of uses by rightsholders. Both systems can also enable rightsholders to "claim" unauthorized material in order to monetize views. Overall, both can give the rightsholder control over content that goes well beyond unfiltered systems. Rightsholders, predictably, tend to approve of filtering options. One went further, and described filtering (in this case, via the default Audible Magic) as necessary for responsible OSPs.

Most OSPs objected strongly to the possibility of creeping requirements in this area. Their concerns fell into two broad categories: the cost of developing or licensing tools; and worries that filtering systems could harm user expression, especially *ex ante* blocking features that could restrain fair or other legal uses prior to publication. None felt that they could match Google's investment in this area, and several viewed the required investment to build or license filtering tools as a barrier to entry that would further consolidate the positions of the major players. Community and UGC sites, especially, described filtering

Many OSPs viewed filtering with considerable suspicion. Community and UGC sites, especially, described filtering as a danger to the free speech norms that animate their services.

[165] *See supra* note 180 and accompanying text.

as a danger to the free speech norms that animate their services. They expressed concerns that such systems could hand the definition of the boundaries of free speech over to rightsholders and into terms of service relationships where fair use protections do not apply. They worried that—while it may be reasonably suited to finding full length movie uploads or songs—filtering does a poor job of parsing all the contextual uses that shape fair use.

When asked about this issue, some rightsholders (though not all) described attempting to limit problems by actively managing content, using the tools provided by filtering systems, rather than engaging in simple blocking. They considered the benefits of this more labor-intensive approach high enough to justify human investment in monitoring user-contested matches. One large rightsholder described Content ID management as a primary focus of enforcement efforts, sufficiently important to warrant a dedicated internal team. As he characterized it, the team works to block obviously infringing uses that compete with existing distribution channels, monetize border cases, and permit probable fair use and fan uses without monetization—with the latter threshold set high enough to permit a range of partial, re-contextualized, and transformative uses. This reflects the nuanced approach some major rightsholders take in deciding what will be considered "tolerated use." But it also reflects a central criticism of filtering systems that support "tolerated use"[166] models: the decision whether or not the use will be allowed is left to the rightsholder rather than the underlying legal rules.

The relationship between filtering systems and the rules of copyright law varies considerably from system to system. In principle, filtering systems do not replace the DMCA or other statutory protections for users; rather, they create an additional layer of private adjudication in which initial determinations about fair use or other limitations and exceptions to copyright are made before the publication of the work. Following this, one OSP broke with the general view, arguing that filtering is also a potentially positive response to the imbalances of automated DMCA practice: not a means of taking more content down, but of keeping content up—of creating a space for negotiation between rightsholder, OSP, and user that isn't afforded under conditions of mass automated sending, where OSP liability dictates takedown of nearly all requests.

In the case of YouTube, these determinations are passed back to the rightsholder. In the case of Vimeo's newer Copyright Match system (which uses Audible Magic fingerprinting technology), Vimeo retains responsibility for blocking or passing through flagged content.[167] Both processes permit appeals by users in the event that content is blocked, after which—in theory—disputes revert to the statutory notice and takedown process and eventually to any legal action the rightsholder may choose to bring. In practice, these forms of recourse also vary greatly, are subject to OSP terms of service, and most importantly, are subject to no legal requirements that they be available at all.

Filtering is still the province of relatively few OSPs with resources and, generally, in the contested music and video spaces.[168] But many of the OSPs among our respondents worried about the expansion of filtering practices beyond the music and video sectors. Others raised

[166] *See* Tim Wu, *American Lawbreaking,* SLATE (Oct. 16, 2007), http://www.slate.com/articles/news_and_politics/jurisprudence/features/2007/american_lawbreaking/tolerated_use_the_copyright_problem.html.

[167] *Copyright Match*, VIMEO, https://vimeo.com/help/faq/legal-stuff/copyright-match (last visited Feb. 5, 2016).

[168] Getty Images has also entered the filtering market with its ImageIRC technology. Bill Rosenblatt, Getty Reaches Image License Deal with Pinterest, COPYRIGHT AND TECHNOLOGY, (Oct. 28, 2013), http://copyrightandtechnology.com/2013/10/28/getty-images-reaches-image-license-deal-with-pinterest/.

the concern that strict filtering requirements will simply drive users to competing sites—including those incorporated in jurisdictions with less exposure to rightsholder pressure. Because the differentiators between many major services are minor, user protections are viewed by several OSPs as important differentiators among services, and key to building site loyalty.

iii. Staydown

Rightsholder groups have long favored legal rules that move beyond requiring OSPs to remove a targeted file in response to a notice to also using hash matching and fingerprinting technologies to block any future attempts to post new copies of the file. This process is often described as "staydown." Staydown requirements would mean that, once rightsholders have identified content on an OSP's platform as infringing, the OSP must prevent the reappearance of that content on their site. As applied to search, RIAA chairman Cary Sherman described the goal as to ensure "that when links to content are taken down, the same content on the same site is not continuously re-indexed when repopulated by the pirate site, rendering the takedown process useless."[169] Among OSPs that use filtering technologies, a form of staydown can be achieved—at least for exact or "fingerprinted" matches—when rightsholders put rules in place that prevent any uploads to an OSP's site of new files matching source files.

For OSPs, staydown represents the worst form of filtering systems that risk mistaken, automated quelling of user expression.

While staydown garners widespread support among rightsholders, OSPs described it in highly negative terms. For OSPs, staydown represents the worst form of filtering systems that risk mistaken, automated quelling of user expression. Indeed, several of our rightsholder respondents expressed a more nuanced stance on the issue, often drawing a distinction between the appropriateness of staydown for exact or inexact matches. For example, one rightsholder respondent acknowledged that technology is not a substitute for human review of fair use questions, and suggested that technological systems should be paired with policies that allow users to object to a match and for the content to be easily reinstated. However, this same respondent favors systems that put up barriers to reposting content identical—i.e., identified through traditional hash matching rather than looser fingerprinting algorithms—to what has already been taken down. While this approach cannot prevent mistaken takedowns where material is licensed or where the context allows fair use of an exactly matching file, it can limit mistaken takedowns overall. The rightsholder considered this tradeoff reasonable.

iv. Side Agreements

A move to side agreements that supplement or circumvent the basic notice and takedown framework is another developing shift in online copyright enforcement. The most prominent was the Copyright Alert System ("CAS"), which was adopted by some connectivity providers following intense pressure from rightsholder groups to adopt explicit termination policies

[169] Ernesto, *RIAA Wants Google to End Piracy "Whack-a-Mole,"* TorrentFreak (March 14, 2014), https://torrentfreak.com/riaa-wants-google-end-piracy-whack-mole-140314/ (quoting RIAA chairman Cary Sherman).

for users—including highly contentious "three strikes" termination policies—as part of their required "repeat infringer" policies.[170]

In 2011, five of the major ISPs announced a deal with several rightsholders groups to adopt and standardize procedures for warning and—after some number of warning notices—sanctioning repeat infringers.[171] The resulting CAS required ISPs to pass "strike" notices identifying allegedly infringing material distributed via peer-to-peer file exchanges on to the targeted users. Users who continued allegedly infringing activities got up to six "strike" notices, each with escalating consequences but—importantly—stopping short of termination, which remained at the discretion of the ISP. Each party got some of what it wanted from the CAS deal: Rightsholders persuaded connectivity services to create a notice process, sanctions, and—potentially—a method for establishing an evidentiary trail should rightsholders choose to take a dispute to court. ISPs deflected pressure for "three-strikes" termination rules and also received caps on the total number of notices that could be sent under the CAS, ameliorating the concern that they could be drowned in automated notices from the rightsholder CAS signatories.

The CAS was in place during this study and lasted until it was disbanded in early 2017. For some ISPs, it mitigated the floods of DMCA notices they were receiving before the agreement. According to several respondents, under CAS participating ISPs no longer received DMCA notices from the Motion Picture Association of America ("MPAA"), RIAA, and their partnering groups. One ISP respondent described the agreement as superseding section 512 and generally reducing the number of headaches associated with managing notices. Going from 1 million to around 100,000 notices per year, he observed, makes a big difference. Although CAS reported sending 1.3 million notices to targeted users in its first ten months of operation,[172] these numbers are around an order of magnitude less than many ISPs received prior to the agreement (under the agreement, however, these numbers were scheduled to double in 2014).[173]

While headaches did still arise—ISPs still reported receiving 512(a) "notices" from non-CAS rightsholders, sometimes very large numbers of them, and rightsholders were not necessarily pleased with the lack of termination—CAS appeared to have some success for both rightsholders desiring to get the word out to users about infringement and the connectivity providers that found themselves caught in the midst of the peer-to-peer file-sharing storm.

We take up other developing and proposed formal agreements—for standardized practices, for tertiary providers such as ad and payments networks—below and in the analysis and

---

[170] By the mid-2000s, major rightsholder groups had begun to campaign for legislation or regulations that could compel ISPs to play a more active role in IP enforcement on their networks. Repeat infringer termination requirements played a large role in these plans—generally under the name "graduated response," which describes a process of escalating notifications and then penalties against users. Graduated response laws were passed and implemented in several countries (including France, New Zealand, Taiwan, South Korea), but the strategy proved more controversial in the US and efforts to include "three strikes" measures in the FCC's National Broadband Plan failed. For a good international discussion, see Rebecca Giblin, *Evaluating Graduated Response*, 37 COLUM. J. L & ARTS 147 (2014), http://ssrn.com/abstract=2322516; *see also* GRADUATED RESPONSE, http://graduatedresponse.org (last visited Feb. 5, 2016).

[171] See CTR. FOR COPYRIGHT INFO., MEMORANDUM OF UNDERSTANDING (2011), http://www.copyrightinformation.org/wp-content/uploads/2013/02/Memorandum-of-Understanding.pdf.

[172] CTR. FOR COPYRIGHT INFO., THE COPYRIGHT ALERT SYSTEM: PHASE ONE AND BEYOND 1 (2014), http://www.copyrightinformation.org/wp-content/uploads/2014/05/Phase-One-And_Beyond.pdf.

[173] *Id.*

recommendations in Section V. Importantly, formal agreements only are the most visible elements of a broader process of norm setting that passes through conflicts, negotiations, business deals, and actual or threatened lawsuits between OSPs and rightsholder groups. We discuss some examples of these types of processes in the case studies below, and take up the underpinning legal rules that provide leverage in negotiations in the Section V analysis and recommendations.

2. "Para DMCA" Measures: Site Blocking and Tertiary Providers

Overall, rightsholders described automated measures, together with education and strategic deployment of takedown, as bringing some meaningful success in enforcing copyright on DMCA-compliant sites. In the United States and European markets, in particular, rightsholders attribute part of their success in encouraging licensed use to shifting attitudes among music consumers. Rightsholders expressed great frustration, however, with the resiliency of extra-territorial file sharing and streaming sites. While some of these sites comply to some extent with notice and takedown, others do not. They are generally beyond US jurisdiction and therefore not nearly as vulnerable to legal threats. The most challenging employ what rightsholders described as "hardcore institutional models built on piracy."

Rightsholders' attempts to cut off the visibility, financing, and access to infrastructure of these sites underpin a good portion of their strategies to target providers that in some way engage with file sharing sites, even indirectly. Here the list includes much of the service infrastructure of the contemporary Internet: ISPs, domain name registrars, ad networks, payment services and search engines, which provide visibility to file sharing sites in the course of indexing the larger web. While ISPs and search providers are covered by section 512, targeting domain name registrars, ad networks, and other services that are unlikely to incur secondary liability moves into "Para DMCA" measures, which may borrow the structure of notice and takedown without the DMCA's procedural framework or protections. Other measures go further still, such as ISP-level site blocking.

a. Site Blocking

Site blocking entails the implementation of ISP-level denial of access to the Internet domains of infringing sites. For non-compliant sites—including a large number of file sharing and file locker sites outside the US—site blocking has been a policy goal for some large rightsholders. Site blocking would extend well beyond even OSP-level *ex ante* filtering mechanisms to block access to an entire Internet domain deemed "dedicated to infringement"—a much broader response than notice and takedown or even fully litigated injunctions, which are subject to careful tailoring.

Accordingly, it is a highly controversial concept, with ISPs, other OSPs, and user groups uniformly opposed on both policy and technical grounds, and a split amongst rightsholders.[174] Several rightsholders and agents in our interviews opposed site blocking, and some major

---

[174] *See, e.g.*, INTERNET SOC'Y., INTERNET SOCIETY PERSPECTIVES ON DOMAIN NAME SYSTEM (DNS) FILTERING (2012), http://www.internetsociety.org/sites/default/files/pdf/dns-filtering_20110915.pdf; Paul Vixie, *Refusing Refused*, CIRCLEID (Jan 11, 2012, 6:41 PM), http://www.circleid.com/posts/20120111_refusing_refused_for_sopa_pipa/.

notice senders publicly withdrew support of the Stop Online Piracy Act ("SOPA") over concerns about the site blocking mechanisms it would have established.[175]

As we explore further in the Section V analysis, the US controversy over site blocking reflected traditional US concerns about government-directed censorship, as someone would need to vet and maintain a blacklist of sites to block.[176] In interviews, rightsholders were not eager to be directly associated with blacklists, or to take on the cost of legal challenges, potential liability, and public outcry that could occur in response to decisions to blacklist specific sites. Some pointed approvingly to blacklists assembled by third parties as a possible solution. So far, however, these exemplify more than solve the problem of controversial identification, and none have gained traction. The largest effort to date—advertising group Group M's blacklist of more than 2000 "pirate sites," assembled in 2011 with rightsholder input—included numerous non-infringing sites and edge cases that would almost certainly generate legal challenges.[177] Existing lists, such as the MPAA's annual "Notorious Markets" report, are thus far used in only an advisory capacity as input into the United States Trade Representative's Special 301 process for pressuring foreign countries on enforcement.[178]

b. Privately Agreed "Best Practices" for Tertiary Intermediaries

The second Para DMCA measure expands enforcement efforts to tertiary players beyond the connectivity, hosting, and search OSPs covered by the DMCA. As discussed above in Section II.B.1, a recent series of government-brokered private "best practices" agreements extend DMCA-like notice-based practices to providers—including advertising and payment providers—that are one, two, or more steps removed from the alleged infringement.

These emerging notice systems represent a logical extension of "follow the money" strategies for combating infringement but—in the absence of a statutory framework or clear jurisprudence on many of these arrangements—OSPs expressed concern that they expand the potential for enforcement with little to guarantee due process. When we conducted our study, these approaches were still in fairly early stages, and affected OSPs did not yet have much experience with them. In interviews, a number of OSP respondents viewed the diffusion of DMCA-like processes across the Internet economy with apprehension, even as they noted that expanding a familiar process to other contexts was the easy thing to do. Several worried that it magnifies the power of a flawed process and increases the potential for abuse. Several thought that it empowers "junk" notice senders and further erodes the procedural protections of section 512.

---

[175] *See, e.g.*, Sean Gallagher, *Even the Business Software Alliance Now Backpedaling on SOPA Support*, ARS TECHNICA (Nov. 21, 2011, 12:45 PM), http://arstechnica.com/tech-policy/2011/11/even-the-business-software-alliance-now-backpedaling-on-sopa-support/. The split went public with the Business Software Alliance's withdrawal of support for SOPA in late 2011, based largely on reservations about the site blocking provisions. *Id.*

[176] On public attitudes toward government and private sector roles filtering and site blocking, see JOE KARAGANIS & LENNART RENKEMA, THE AM. ASSEMBLY, COPY CULTURE IN THE US AND GERMANY (2013), http://piracy.americanassembly.org/wp-content/uploads/2013/01/Copy-Culture.pdf.

[177] Ernesto, *BitTorrent.org and Archive.org Blacklisted as Pirate Sites by Major Advertiser*, TORRENTFREAK (June 10, 2011), https://torrentfreak.com/bittorrent-com-and-archive-org-blacklisted-as-pirate-sites-110610/. In 2014, the City of London Police solved this problem by keeping their new advertising site blacklist private. Stuart Dredge, *Forget Suing Filesharers: In 2014, Anti-Piracy Efforts Follow the Money*, GUARDIAN (Apr. 2, 2014), http://www.theguardian.com/technology/2014/apr/02/infringing-websites-list-anti-piracy.

[178] *See United States Trade Representative*, 2015 SPECIAL 301 REPORT (2015), https://ustr.gov/sites/default/files/2015-Special-301-Report-FINAL.pdf.

One OSP also pointed out that the move to mobile platforms creates additional complexity for targeting ads. The SoundLocker case study[179] below provides a striking example of how bringing ad networks into enforcement could begin to reverberate back to DMCA-compliant OSPs in the interconnected online ecosystem. As these services consolidate into larger interconnected platforms—sometimes including hosted content, advertising platforms, mobile platforms, and more under the control of a single entity—notices have begun to circulate across them in ways that create a more generalized and unpredictable climate of liability. Rightsholders, however, viewed the same developments as an extension of "follow the money" strategies for attacking large-scale infringement. We take up these themes in Section V discussion, below.

E. COMPLIANCE, COMPETITION, AND MARKET POWER

Several of the small OSPs in our study viewed the shift toward automation and DMCA Plus enforcement as a source of competitive advantage for larger OSPs, which could better bear the costs of developing and managing these systems and exercise more leverage in negotiations with rightsholders. Small OSP players that had been targeted by a large volume of notices for whatever reason described struggling to manage the substantial costs of compliance, giving larger and better-resourced players an advantage. Much of this advantage resides not just in the cost of systems but also in their cumulative cost as demands increase. The struggle increased further if pressure to implement DMCA Plus measures arose. Automated notice handling does not obviate the need for human review. Content filtering systems do not obviate the need for automated notice handling. Even if they are off-the-shelf, they must be integrated into OSPs systems. And rightsholders vary in their requests, pushing OSPs to run parallel systems to manage different enforcement demands.

In some striking cases, it appears that the vulnerability of smaller OSPs to the costs of implementing large-scale notice and takedown systems and adopting expensive DMCA Plus practices can police market entry, success, and competition. Those without sufficient resources to build or license automated systems described being in precarious positions, at risk of being priced out of the market by better-resourced competition if floods of notices or DMCA Plus requirements were to arrive.

Most of the conversations with OSPs focused on content filtering. None viewed homegrown development of such systems as a viable option. Several described filtering as a major competitive advantage for larger services, which can afford to either license or develop fingerprinting and filtering technologies (or both). All were familiar with estimates of the cost of development of Google's Content ID system, which begin at $60 million.[180] Commercial fingerprinting and filtering services, such as Audible Magic and Vobile, do not publicly release pricing. But we can guess at the ballpark: one medium-sized file hosting service reported that its license for Audible Magic filtering cost $10,000-12,000 per month in 2011 (though this provider was later able to negotiate a reduced rate based on the amount of content flagged through the system). Another estimated that Audible Magic cost its service roughly $25,000 per month. OSPs noted that the licensing fees are just the beginning. Filtering systems, several OSPs noted, are not turnkey services. They require integration with existing systems and upkeep as the OSP takes on new mediation roles between rightsholder and user (such as tracking and managing user appeals).

---

[179] SoundLocker is a pseudonym.

[180] *Hearing on Section 512 of Title 17, supra* note 86, at 49 (testimony of Katherine Oyama, Senior Copyright Policy Counsel, Google Inc.). Estimates may vary in the types of costs they take into account. One well-informed respondent suggested that the full cost was likely several times higher.

Several OSPs expressed concern about the diverging interests of small and large OSPs in this environment. One music service OSP noted that for most of the time that DMCA notice and takedown has been in operation, small and large OSPs' interests in protecting the notice and takedown system were aligned. However, as well-resourced players have implemented filtering systems, this provider worried that "we will see a world where [these large incumbents] start to close the gate behind them" and advocate for DMCA Plus measures. OSP interests are diverging, this respondent stated, as the incumbent OSPs' interest shift to align with rightsholders. This OSP felt that if filtering becomes the "law of the land," then the higher costs of entry will undermine entrepreneurialism and small service providers will not be able to compete. These concerns were exacerbated by the limited market competition for non-bespoke fingerprinting and filtering systems. As noted above, Audible Magic dominates the market. Other services, such as Vobile for audiovisual content, have much smaller market share and visibility.

Not all of this advantage derives from cost of development or adoption. One video hosting platform respondent explained that Google has a significant competitive advantage because of the integration of Content ID with Google's advertising services. The respondent considered Content ID as providing a significant competitive advantage for YouTube because it enables content owners to monetize content that is posted by users and hosted on the platform. This OSP provided examples of video content that is flagged by matching technology—including fan vids, remix videos, and wedding videos that include songs—that is unlikely to be hosted on its site because, unlike YouTube, it does not have Content ID-like system that allows in-video advertising revenue shares with rightsholders. This OSP speculated that, in the long term, when users seek out a platform to share videos that include copyrighted content—regardless of whether the use is fair or otherwise allowed—they will just use YouTube in order to avoid takedown. As other OSPs cannot come near affording what they expect Content ID cost, those with audio and visual offerings considered Google to have a profound competitive advantage.[181]

Rightsholder respondents were generally unsympathetic to the view that enforcement requirements might create significant barriers to entry. One tartly described it as a "pirate theory of innovation," whereby services try to become large enough through infringement to eventually force rightsholders into licensing agreements.[182] The target of this comment was YouTube, but the criticism encompassed a wide range of other cases, from successful Chinese search and social media services to unsuccessful music and cloud storage websites.

Rightsholder respondents were generally unsympathetic to the "pirate theory of innovation," whereby services try to become large enough through infringement to eventually force rightsholders into licensing agreements.

---

[181] Though it did not come up in interviews, Vobile is positioning itself as a "turn-key" solution to provide Content ID style monetization to other OSPs. *Monetization – Content Monetization*, VOBILE, http://www.vobileinc.com/monetization (last visited Feb. 5, 2016). It remains to be seen other options will arise, and if this will be sufficient to diminish Content ID's competitive advantage.

[182] Invoking the "pirate theory" of innovation locates the current debate in a long-running scholarly narrative about how new media producers historically challenge the control over distribution of older media incumbents, and then become incumbents, themselves. ADRIAN JOHNS, PIRACY: THE INTELLECTUAL PROPERTY WARS FROM GUTENBERG TO GATES (2010); SOC. SCI. RESEARCH COUNCIL, MEDIA PIRACY IN EMERGING ECONOMIES (Joe Karaganis ed., 2011), http://piracy.americanassembly.org/the-report/.

1. Case Study: SoundLocker

The experience of the music service SoundLocker[183] illustrates some of the potential market pressures faced by new music or video platforms. It provides a deeper example of how pressure to implement DMCA Plus and Para DMCA measures can unfold, and what it can cost a start-up company. SoundLocker provides a platform for artists to upload and distribute their own music. Though it was not the intention, like other sites that rely on user attestations of ownership, SoundLocker also began to attract unauthorized music uploads.

As the site grew, it attracted the attention of the RIAA, which began to send warning letters and DMCA notices to SoundLocker—coupled with matching DMCA notices to Google Web Search. According to SoundLocker, it complied to the letter with the RIAA notices, removing everything it could identify regardless of doubts about the validity of some takedown requests. The RIAA's notices to Google Web Search, however, resulted in the suspension of SoundLocker's Adsense account in 2011.

Because SoundLocker's revenue was derived from advertising, it reached out to the RIAA for support in getting its Adsense account reinstated, asking for the RIAA "wish list" of terms that would put SoundLocker in good standing. The RIAA's responding list included an array of DMCA Plus measures we have already discussed at length, including unlimited direct administrative access for takedowns with no review by SoundLocker and the adoption of Audible Magic filtering software. SoundLocker implemented the RIAA wish list in early 2012, though with considerable hesitation as its contract with Audible Magic represented one-third of its advertising revenue.

For the next year, neither SoundLocker nor Google received significant numbers of notices related to content on the SoundLocker site, and none from the RIAA. But by early 2013, SoundLocker had to cut costs. It approached Audible Magic for a reduced rate and was refused. It then approached the RIAA to see if it would subsidize the use of Audible Magic. The RIAA said no. As SoundLocker's CEO put it: "I had concluded that Audible Magic was just an RIAA tax. We were under no legal obligation to use it. Given the cost, we decided that we could do DMCA compliance on our own." SoundLocker dropped Audible Magic in mid-2013.

Six weeks later, the RIAA contacted SoundLocker to ask if it was still using the Audible Magic service. SoundLocker replied that it was not. A couple of weeks after that exchange, Google began to receive tens of thousands of SoundLocker takedown requests from the RIAA—peaking at 18,000 per week in September. According to SoundLocker's CEO, almost all the requests were "obtuse or grossly inaccurate," pointing to dead links, non-existent links, previously removed links, and numerous pages of search results (rather than specific links). In September, Google canceled SoundLocker's Adsense account for a second time—this time with $400,000 in undisbursed revenue in SoundLocker's account.

Round two looked much like round one. SoundLocker went back to Audible Magic to negotiate for a reduced rate—this time on threat of shutting down its own site. Audible Magic agreed to a steep discount and SoundLocker reinstating the service at a fraction of the earlier cost. SoundLocker went back to the RIAA to protest the bad takedowns, announce the renewed Audible Magic contract, and request RIAA support in restoring the Adsense account. The RIAA agreed and conversations with Google were reopened, but by this

[183] As noted above, SoundLocker is a pseudonym.

time Google had received a wave of new notices from REOs, primarily Topple Track and Audiolock.net, which represented small labels. Google rejected the request.

And so it continued. SoundLocker threatened the new senders with lawsuits for misrepresentation. They provided affidavits promising to stop sending notices. Google did not respond.

For SoundLocker, the experience became a Catch 22. Adsense was a critical service for SoundLocker: it paid double the rates of other services, and didn't subject users to tool bar installers or other scamware. But the process of dispute resolution around the service was opaque and arbitrary. SoundLocker's CEO at first thought he had to please the RIAA, but found that in practice any rights group had blocking power. He described Google as unresponsive during much of the dispute period. Throughout, the RIAA, the International Federation of the Phonographic Industry ("IFPI"), and many other rights enforcement groups had direct takedown access to the SoundLocker service. All the gatekeepers, meanwhile, were invested in competing music services—the RIAA's members in Spotify, Google in YouTube and Google Play.

Today, SoundLocker operates without Audible Magic, at a fraction of its peak size. According to its CEO, SoundLocker no longer receives many notices. It won some peace from the RIAA and the smaller automated senders. But has also been beaten by costs and demands that quickly grew, that extend beyond the DMCA's statutory requirements, that vested control in third parties, and that had no clear process for remediating mistakes. As SoundLocker's CEO put it: "I'm an American citizen. We didn't want to run a rogue site. We didn't want pirated Lady Gaga songs. We tried to do everything right." But in the end, "I didn't have 10 million dollars to file a lawsuit."

F. THE ROLE OF SEARCH SERVICES

Search services occupy a slightly odd place in the section 512 landscape. Links to content hosted elsewhere are generally not infringing in and of themselves.[184] Yet findings of secondary liability for linking are possible.[185] Section 512 accordingly provides a notice and takedown process for "information location" tools such as search engines that affords them

[184] Hyperlinking does not itself constitute direct copyright infringement because there is no copying. *See, e.g.*, Online Policy Grp. v. Diebold, Inc., 337 F. Supp. 2d 1195, 1202 (N.D. Cal. 2004); Ticketmaster Corp. v. Tickets.Com, Inc., No. CV 99-7654 HLH(BQRX), 2000 WL 525390, at *2 (C.D. Cal. Mar. 27, 2000); *see also* Perfect 10, Inc. v. Amazon.com, Inc., 508 F.3d 1146, 1161 (9th Cir. 2007) (holding that providing HTML instructions that direct a user's browser to a website housing copyrighted images "does not constitute direct infringement of the copyright owner's display rights" because "[p]roviding HTML instructions is not equivalent to showing a copy"); Pearson Educ., Inc. v. Ishayev, 963 F. Supp. 2d 239, 251 (S.D.N.Y. 2013) (stating that because hyperlinks do not contain the copyrighted content, forwarding them does not directly infringe on a copyright owner's exclusive rights).

[185] Although hyperlinking per se does not constitute direct copyright infringement, "in some instances there may be a tenable claim of contributory or vicarious liability.". *Diebold*, 337 F. Supp. 2d at 1202 n. 12. Contributory copyright infringement occurs by "intentionally inducing or encouraging direct infringement" of a copyrighted work. Metro-Goldwyn-Mayer Studios, Inc. v. Grokster, Ltd., 545 U.S. 913, 914 (2005). In *Intellectual Reserve, Inc. v. Utah Lighthouse Ministry, Inc.*, the court granted a preliminary injunction holding that the plaintiff was likely to succeed on the merits of the case for contributory infringement. 75 F. Supp. 2d 1290, 1295 (1999). The court held that the defendant in the case actively encouraged infringement on the plaintiff's copyright when, after being ordered to remove the content from its website, it posted links to three websites containing the plaintiff's content and actively encouraged the downloading, copying, and posting of the material on other websites. *Id.* at 1294-1295.

safe harbor protection. Indeed, in interviews and surveys, we found that search engines play a unique role in takedown.

Although the DMCA anticipated that information location services might play an important role in copyright enforcement, the notice and takedown process barely touched them for a decade. Prior to 2009, Google Web Search received, at most, a few hundred requests per year, and more often a few dozen. A few years later that number was in the hundreds of millions. Rightsholders had begun to prioritize the role of search in the discovery of infringing content, and automation had given them a means of exercising their rights on a scale commensurate with the indexing function of the major search engines. Rightsholder attention shifted accordingly. As MPAA Chairman Chris Dodd put it in 2013, search engines "bear responsibility for introducing people to infringing content, even people who aren't actively looking for it."[186]

As a practical matter, search is potentially an especially important target for rightsholders: if infringing material is less discoverable, then rightsholders may be more successful in stemming the tide, especially for wide-scale infringement, than with notices to hosts. In practice, however, both OSP and rightsholder respondents agreed that targeting search results has a limited effect on determined infringement. Those who want to reach unauthorized files often travel to the source directly or find them through channels other than traditional search services.

Given this, rightsholders described submitting takedown notices to search services as an incomplete—though useful within a narrow band—tool for addressing wide-scale infringement. Given these limitations, rightsholders have pushed search services to adopt strategies above and beyond traditional notice and takedown—for example, "de-prioritizing" or delisting troublesome sites.

In this vein, most rightsholders emphasized their desire for search services to take a more proactive role in reducing the occurrence of links to infringing content. For rightsholders, the efficacy of search engine enforcement is measured not by specific takedowns, but by the relative ease of availability of infringing materials. As one rightsholder put it "If you type in a title and within the first page can find illegitimate content, there is still a problem." How such content is kept off the first page of search returns is viewed as less important: "whether accomplished by de-prioritizing or delisting; that's an academic debate, really." The automated notice and takedown process, in this context, serves mostly to push pirate sites down in rank so that they no longer appear in the first pages of search returns.

One sender stated that its goal in targeting search results is to prevent unauthorized channels from appearing on the first page of search returns: "whether accomplished by de-prioritizing or delisting; that's an academic debate, really."

One rightsholder said that, in an ideal world, infringing sites would be completely removed from the search engine index, but in practice the rightsholder hopes that targeting the sites with notices will prompt search engines to push them down so they no longer appear in the first two pages of search returns. This can work because some search providers factor the number of takedown notices they receive for a site into their search ranking algorithms,

[186] Ted Johnson, *Showbiz Lobby Puts Capitol Hill Pressure on Google to Take Action on Piracy*, VARIETY (Sept. 18, 2013, 9:58 AM), http://variety.com/2013/digital/news/google-pressured-by-hollywood-to-do-more-to-fight-piracy-1200616306/ (quoting Dodd).

demoting sites that are the targets of notices. One rightsholder explained that the possibility of site demotion is an important part of its calculus as it decides which sites to target—if a site has a low ranking to start with, but starts to move higher up in a provider's search index, it will start sending notices targeting that site to the search provider. However, this same rightsholder said that if a site is already "hugely popular" and shows up high in the index, notice sending does not seem to demote that site's ranking. This was echoed by another rightsholder, who also said that strategies of sending notices to search targeting sites that are "major indexes for pirated content" had no effect on those sites rankings.

The major rightsholders have pushed for other measures to make finding unauthorized file sharing sites more difficult, including, search term filtering, blocking auto-complete for designated terms, rating systems for "good" and "bad" sites, and—at the limit—blocking sites entirely from search results.

Search services generally described being protective of their indexing and search algorithms, which they viewed as part of the infrastructure of the open Internet. Accordingly, they were reluctant to accede to some changes. As in other areas, however, there is room for negotiation. One search respondent described his company's history of experimentation with "reasonable and effective" measures beyond notice and takedown, but viewed most of the strategies proposed by rightsholders as ineffective, and therefore certain to be followed by more demands if adopted. Others reach other conclusions. Google, for example, has adopted autocomplete restrictions[187] and search result demotion based on the number of takedown notices a site receives.[188]

Search providers also argued that they were poor targets for addressing piracy, because users do not rely on their services to find infringing content to any great degree.[189] Several senders also acknowledged that search does not produce a significant amount of traffic to the most popular pirate sites. They recognized that consumers locate pirate sites in a variety of ways and that once a pirate site is located, consumers will go directly to that site. One rightsholder respondent noted that it had abandoned mass sending of notices to search in favor of narrower approaches that exploit "weaknesses" in the pirate ecosystem—such as the less frequent search indexing and lower community resilience of many non-English-language, small community sites.

Still, rightsholders considered sending takedown notices to search services as one useful tactic in a broader strategy. Though rightsholders generally agreed that targeting search results is, as one put it, not a "silver bullet," all considered search takedown important. Search is seen

---

[187] Kent Walker, *Making Copyright Work Better Online: A Progress Report*, GOOGLE PUBLIC POLICY BLOG (Sept. 2, 2011), http://googlepublicpolicy.blogspot.com/2011/09/making-copyright-work-better-online.html (stating that Google filters terms closely associated with infringement from Google Autocomplete).

[188] GOOGLE, HOW GOOGLE FIGHTS PIRACY, *infra* note 193 (stating that Google factors in the number of valid copyright removal notices it receives for a given site as one signal among hundreds that it takes into account when ranking search results). *See also* Katherine Oyama, *Continued Progress on Fighting Piracy*, GOOGLE PUBLIC POLICY BLOG (Oct. 17, 2014), http://googlepublicpolicy.blogspot.com/2014/10/continued-progress-on-fighting-piracy.html (stating that Google refined the signal it uses to downrank sites for which it receives a large number of valid DMCA notices and expects that this will visibly affect the rankings of some of the most notorious sites).

[189] *Cf.* MILLWARD BROWN DIG., UNDERSTANDING THE ROLE OF SEARCH IN ONLINE PIRACY (2013), http://www.mpaa.org/wp-content/uploads/2014/03/Understanding-the-role-of-search-in-online-piracy.pdf; Mike Masnick, *Data Shows: Removing 'Rogue Sites' From Search Won't Make Much Of A Difference*, TECHDIRT (Nov. 30, 2011, 12:50 PM), https://www.techdirt.com/articles/20111130/05022316931/data-shows-removing-rogue-sites-search-wont-make-much-difference.shtml.

as an important target at crucial times, such as just before or just after the release of new entertainment properties. Moreover, several senders described sending notices to the hosting site and to Google Web Search in parallel. (Other search providers also receive notices as part of this strategy, but Google Web Search came up most often in conversations with rightsholders, and other search providers reported receiving fewer requests than the number shown in Google's Transparency Report.) Rightsholders described sending notices to search providers as a part of an attempt to "attack piracy from all directions," and explained that targeting search is a part of their overall strategy to, at minimum, make infringing content "a little bit harder to find."

Rightsholders described targeting search as a part of an attempt to "attack piracy from all directions," and explained that targeting search is a part of their overall strategy to, at minimum, make infringing content "a little bit harder to find."

However deep rightsholder skepticism about the value of search notices runs, it has not visibly changed the trend line notices to for Google Web Search. According to Transparency Report data, takedown requests continue to climb, reaching around 17 million per week in late 2015 and on pace for around a billion in 2016. As the cost of sending requests falls, and as most rightsholders take a "try everything" approach to enforcement, there appears to be little reason for these trends to change.

1. Case Study: The Disproportionate Role of Google Web Search

The disproportionate role of Google Web Search ("GWS") in notice and takedown deserves closer attention, not least because its actions have repercussions throughout the OSP ecosystem. Respondents at search engines other than Google generally reported receiving dramatically fewer notices than GWS (though still significant numbers as compared to other respondents), and when rightsholders described duplicating notices to host sites with parallel notices to search, they were usually referring to Google's search service. The question of how Google became a magnet for attention is usually answered in terms of its dominant position in search and its active role in debates over enforcement, which frequently put it at odds with rightsholder groups.[190] While there is clearly some truth in these explanations, rightsholders and OSPs tell a more complicated story, with several contributing factors.

As automated sending systems became common, a number of rightsholders began to send duplicate takedown requests, one to the primary file sharing site or host, and one to GWS for takedown of the link to the material.[191] In interviews, several sender respondents also reported that they practice rough parity in sending to file sharing sites and GWS, though these practices vary. One noted that it does not bother sending a notice to file sharing sites that do not respond to notices, opting instead to send notices only to GWS in this scenario. One rightsholder respondent expressed a more nuanced strategy, reporting that although it

[190] As noted above, its Transparency Report sometimes comes up as well, both as an example of provoking senders and—in the opposite vein—as a public record on which REOs and other senders can showcase their efforts.

[191] One-to-one parity does not appear to be so common at lower levels, where different rights enforcement organizations pursue a wide range of strategies against file sharing sites. Information at this level is quite limited and mostly self-reported. Kickass Torrents reported receiving 278,864 URL takedown requests from rightsholders in 2013. Ernesto, *Obsessed with Google, Copyright Holders Ignore the Actual Pirated Content*, TorrentFreak (Apr. 15, 2013), http://torrentfreak.com/obsessed-with-google-copyright-holders-130415. Google, in the same period, received 1,344,885 requests for removal of Kickass Torrent URLs. *Id.*

used to practice rough parity in notice sending it now "intelligently" targets GWS only when it is likely to be effective in lowering a hosting site's search ranking.

It appears that technological feedback loops, fueled by automation on both sides, have played an important role in so rapidly increasing the numbers of notices to GWS. Google's capacity to receive notices grew in parallel to rightsholders' ability to send them. With the adoption of form-based submission, bulk submission, the adoption of the Automated Copyright Notice System,[192] and trusted sender programs, the cost of sending a notice to Google plummeted. Accordingly, GWS appears to be an attractive target for takedown demands in part because sending notices to it can be a low-cost, high-volume endeavor.

From the perspective of some other OSPs, Google's size, its prominence in the politics of notice and takedown, and its role in litigation, combined with its early adoption of DMCA Plus measures like content filtering on YouTube, trusted sender programs,[193] autocomplete restrictions,[194] and search result demotion,[195] make it a dangerous elephant in the room. It is capable of adopting practices that could move collective perceptions of what is required for good practice, or even for safe harbor protection. When Google adopts DMCA Plus measures, these OSPs see their own practices under threat, as they fear the norm-setting potential of these moves.

Some OSPs view Google as a dangerous elephant in the room, capable of adopting practices that could move the collective boundaries of safe harbor protection.

Some OSP respondents speculated that GWS is an attractive target because Google's Transparency Report provides a rough public metric of takedown efforts. Unlike most OSPs, Google provides a visible record of GWS takedowns when it publishes the Transparency Report. An anonymous staffer from an REO wrote to TorrentFreak in apparent agreement:

> Copyright holders are interested in Google only for its "visual effect." They can "see" how many links are removed so it's easier for removal companies to show the ROI. (it [sic] makes them look like they are achieving something).[196]

This is not wholly convincing, as rightsholders explained to us that they receive reports of takedowns directly from the REOs they hire. However, public metrics could help REOs generate new business. Notices that disappear into cyberlocker or torrent sites offer no such confirmation.

---

[192] Automated Copyright Notice System, http://acns.net (last visited Feb. 5, 2016).

[193] *See* GOOGLE, HOW GOOGLE FIGHTS PIRACY 14 (2014), https://drive.google.com/file/d/0BwxyRPFduTN2NmdYdGdJQnFTeTA/view.

[194] Kent Walker, *Making Copyright Work Better Online: A Progress Report*, GOOGLE PUB. POLICY BLOG (Sept. 2, 2011), http://googlepublicpolicy.blogspot.com/2011/09/making-copyright-work-better-online.html (stating that Google filters terms closely associated with infringement from Google Autocomplete).

[195] GOOGLE, HOW GOOGLE FIGHTS PIRACY, *supra* note 193, at 18 (stating that Google factors in the number of valid copyright removal notices it receives for a given site as one signal among hundreds that it takes into account when ranking search results); *see also* Katherine Oyama, *Continued Progress on Fighting Piracy*, GOOGLE PUB. POLICY BLOG (Oct. 17, 2014), http://googlepublicpolicy.blogspot.com/2014/10/continued-progress-on-fighting-piracy.html (stating that Google refined the signal it uses to down rank sites for which it receives a large number of valid DMCA notices and expects that this will visibly affect the rankings of some of the most notorious sites).

[196] Ernesto, *Obsessed with Google, supra* note 191.

More broadly, public metrics are more politically useful than private metrics. The number of takedowns Google accomplishes in a year can be useful to rightsholders requesting additional remedies against infringement, as they can point to the Transparency Report to show the scale of the problem.[197] Equally, this public metric can be used by Google to show how efficiently it removes infringing material.[198] The flood of notices to Google also clearly targets wider political conversations about enforcement, including federal and state legislative efforts and state-level investigations.[199] The notices are part of a public case that Google bears unique responsibility for enforcement based on its role in facilitating the discovery of unauthorized content.[200]

A more pointed question is whether the notice escalation represents coordinated retaliation against Google for its role in opposing stronger enforcement obligations on OSPs, culminating in the defeat of the Stop Online Piracy Act in 2012. The timing of the escalation outlined by Seng is suggestive.[201] 2011 saw a large 305% increase in notices over 2010.[202] But the banner year was 2012, with a 524% increase overall and a 227% increase April alone, in the immediate aftermath of the withdrawal of the bill.[203] Most of the major rightsholders only began actively targeting Google Web Search during this period, including those who currently send massive numbers of notices: the RIAA and BPI, the major movie studios, and Degban Ltd.—the lead REO for the pornography industry.[204]

In interviews, rightsholder representatives rejected any suggestion of retaliatory motives for the explosion of search takedowns, but all acknowledged that 2011-2012 was the period in which the content industries aggressively tested the potential of using notices to search services to limit access to unauthorized materials. As we describe above, however, some OSPs—notably those whose platform offerings include music—received batch "floods" of notices that they experienced both as retaliatory and as policing market entry. We note that retaliation and legitimate interest in testing the limits of notice and takedown are not mutually exclusive motivations. Overall, evidence points to both.

---

197 *See, e.g.*, Brad Buckles, *One Year, 20 Million Links to Illegal Songs Sent to Google: This Is How It's Supposed to Work?*, RIAA MUSIC NOTES BLOG (May 22, 2013), http://www.riaa.com/blog.php?content_selector=riaa-news-blog&content_selector=riaa-news-blog&blog_selector=One-Year-&news_month_filter=5&news_year_filter=2013.

198 *See, e.g.*, *Hearing on Section 512 of Title 17*, *supra* note 86, at 47 (testimony of Katherine Oyama, Senior Copyright Policy Counsel, Google Inc.).

199 Joe Mullin, *Hollywood v. Goliath: Inside the Aggressive Studio Effort to Bring Google to Heel*, ARS TECHNICA, (Dec. 20, 2014, 4:17 PM), http://arstechnica.com/tech-policy/2014/12/how-hollywood-spurned-by-congress-pressures-states-to-attack-google/.

200 Among many examples, see comments by MPAA Chairman Chris Dodd in 2013: "Search engines bear responsibility for introducing people to infringing content, even people who aren't actively looking for it." *See* Ted Johnson, *Showbiz Lobby Puts Capitol Hill Pressure on Google to Take Action on Piracy*, VARIETY (Sept. 18, 2013, 9:58 AM), http://variety.com/2013/digital/news/google-pressured-by-hollywood-to-do-more-to-fight-piracy-1200616306/ (quoting Dodd).

201 *See* Seng, *supra* note 5, at 389.

202 *Id.*

203 *Id.* at 389-90.

204 *See Transparency Report*, GOOGLE, *supra* note 12 (see the *Owners* subsection). The possibility of such coordination gained credibility with the news of Project Goliath—the studios' 2014 anti-Google program revealed in the Sony document leak. The Project Goliath emails make it clear that targeting Google was an explicit priority for the studios and part of a broader policy agenda to challenge safe harbor protection and establish legal frameworks for site blocking. *See* Russell Brandom, *Project Goliath: Inside Hollywood's Secret War Against Google*, VERGE (Dec.12, 2014), http://www.theverge.com/2014/12/12/7382287/project-goliath.

For all the rightsholders' fear of Google's ability to undermine their control of their products, it also benefits the major rightsholder groups for Google (or some other single provider) to be the dominant provider of discovery services. Because of its dominance, its private agreements can meaningfully affect user activity, set norms for other service providers, and provide a simpler target for both regulatory action and further private negotiation. Google is in much the same position: split between its origins as an index for the open web and its increasingly powerful positions across a range of services, it can police other OSPs' activity by docking search rank and AdSense revenues for OSPs that attract large numbers of notices, and it can wield Content ID to attract revenues and offer rightsholders benefits that other OSPs cannot afford. This is a tough position in that decisions about search rank and AdSense revenue might invite criticism, but also an enviable one in terms of market position for licensed services.

The SoundLocker case study presents a dramatic example of the manner in which market entry, costs, and enforcement expectations can be affected by the decisions of much larger parties, but we heard similar concerns from other OSPs. In the end, Google's place in the notice and takedown ecosystem provides an important example of how litigation, negotiation, and compromises between the most dominant players in the online ecosystem can affect a much broader swath of the system.

G. DISCUSSION: STUDY 1

What conclusions can we draw from this exploration of the section 512 landscape? First is the continuing importance of the section 512 safe harbor and the notice and takedown process. All OSPs affirmed the importance of section 512's safe harbor. Indeed, they so strongly rely upon its ability to limit liability and reduce uncertainty that they consider it foundational to their ability to provide intermediary services. Rightsholders reiterated their oft-stated frustrations with notice and takedown for addressing large-scale Internet infringement, but also described relying on it as a crucial component of policing their copyrights.

Second, the experiences of OSPs with notice and takedown diverge dramatically, as do their notice and takedown practices. While all OSPs rely on the safe harbor, its practical availability is mediated by their relative resources and the scale of enforcement efforts directed at them. As both infringement and takedown have scaled up for some parties, the nature of the DMCA's influence has changed for those dealing with large-scale infringement and automated noticing. Automation has increased the gross day-to-day use of the notice and takedown process for the OSPs that receive large volumes of notices, but the practical importance of any individual notice under the statutory process has diminished for many of these providers. Further, while the statutory requirements have not changed, pressure to adopt stronger DMCA Plus measures like back-end access or filtering has grown for some OSPs, reinforced by litigation, threats of new legislation, and private agreements among stakeholders. Traditional DMCA compliance is therefore an increasingly fragile structure for a subset of OSPs—those vulnerable to the fire hose of automated notices and eroded by the adoption of DMCA Plus measures by some OSPs. The DMCA's safe harbors set up and maintain basic roles, yet under the pressure of Internet-scale policing market power, forbearance, and luck play a growing role in maintaining some OSPs' *de facto* freedom to operate. For these OSPs, notice and takedown can function less as a clear practice that services must implement than as a legal standard that underlies ongoing negotiations between rightsholders and service providers about anti-infringement practices.

Third, a large number of our OSP respondents—indeed, the great majority—are simply not a part of the shift to DMCA Auto or DMCA Plus enforcement measures detailed above.

They do not receive large volumes of notices and do not employ automated review systems; for these OSPs, "classic" notice and takedown remains the norm. For them, takedown works reasonably well, and the safe harbor represents a key protection. They tend to be less visible in public debates about notice and takedown where large rightsholders struggling with Internet-scale infringement and large OSPs managing floods of notices tend to take center stage. Accordingly, they are acutely aware that public debates over notice and takedown may leave their interests aside. They described the political and litigation-based wrangling over the safe harbors with trepidation borne of a concern that relatively few actors' outsized concerns could harm the legal or practical availability of the safe harbor.

Most DMCA Classic OSPs view their status as dynamic, and some as tenuous. These OSPs are very aware that, when dominant OSPs adopt stronger DMCA Plus measures, their actions can acquire the force of new norms that move general perceptions of liability and make wider enforcement agreements more likely. DMCA Classic OSPs maintain that they have neither the resources nor the leverage with rightsholders to operate effectively in a DMCA Plus world.

Players who implement DMCA Plus measures, on the other hand, typically engage in frequent negotiation with rightsholders. One respondent described a constantly evolving dialog with rightsholders regarding measures beyond notice and takedown: "The list of demands [for measures beyond section 512] seems to change about every six months." From his perspective, rightsholder demands should be outcome driven, with solutions left to the OSPs. Rightsholders, in turn, expressed frustration that OSPs' perceived lack of independent action left them specifying technical solutions to infringement. Such disagreements fuel the ongoing conflict around the DMCA. Rightsholders see existing measures fail and conclude that they need more. OSPs see them fail and see a slippery slope.

Fourth, the targets of notices—Internet users—are often lost in this struggle over duties. As a practical matter, the targets of notices are at the mercy of others, and appear to be at risk of being subsumed by rightsholders' more immediate priorities of removing infringing content and of OSPs' need to avoid liability. The main formal protection afforded by the DMCA—the counter notice process—is widely viewed as ineffective, empowering unscrupulous users and subjecting legitimate ones to legal jeopardy. Targets (other than bad-faith, off-shore pirates) were widely considered to lack sufficient information to respond to mistaken or abusive notices. DMCA Plus measures such as filtering can provide more flexible means of resolving disputes but provide only as much due process as terms of service agreements dictate. The substantive problems with notices we report in Study 2 and Study 3 below further suggest that targets are not well-served by section 512.

In Section V, we take up some recommendations to improve notice and takedown for targets, and also suggest further research into their experiences.

Fifth, Study 1 revealed a major lack of transparency around takedown practice, and attendant information asymmetries. OSPs described operating largely in the dark about their competitors' practices and—in DMCA Classic cases, especially—the triggers that might attract rightsholder attention. DMCA Classic OSPs in this "fragile" camp tended to be unsure why they had not been deluged with automated notices.[205] The most common result is that OSPs act conservatively, erring on the side of takedown and over enforcement in order to avoid "bet[ting] the company" on disputes at the peripheries of their core business practices.

Finally, the automation of notice and takedown and its role in pushing OSPs beyond DMCA requirements appears to be a factor in the consolidation of market power by large intermediaries—both in terms of economic costs, as smaller OSPs face expensive requirements, and policy costs as requirements consolidate around the needs and capacities of the larger players. At the limit, automation and pressure to adopt DMCA Plus measures have the potential to significantly affect market entry and competition for online platforms. If automated noticing or pressures to adopt measures like filtering become the norm, the high potential liability for secondary copyright infringement could easily combine with much higher costs of compliance to form a potent barrier to entry for OSPs with fewer resources.[206] This has always been a risk, as notice and takedown compliance has never been (and should not be) costless. But the risk appears to be intensifying as major players like Google consolidate advertising, content discovery, and filtering technologies, and in the process gain competitive advantages in their ability to comply with legal requirements and emerging norms. As we discuss in the Section V analysis and recommendations, this puts additional pressure on policymakers to avoid legal or norm-based changes that would create barriers to entry. The Internet is both the most distributed forum for expression in human history and dependent on private intermediaries to support that expression. Notice and takedown should support, not limit, competition to provide those services.

---

[205] This is not to say that OSPs' concerns are unreasonable or would change based on greater transparency—for some, pressure comes from adult entertainment providers or other senders outside of the group of rightsholders with which we spoke. In the two studies we report on next, we found that while major rightsholders do tend to target the "worst of the worst" via Google Web Search (Study 2), Google Image Search notices are dominated by a small handful of senders, all of whom are outside of the major copyright industries and conflict zones, and most of whom are individual or small-business senders (Study 3). *See supra* Section IV. Together, these senders produced tens of thousands of notices in a six-month period—enough to swamp a small OSP.

[206] For a discussion of stakeholders' views on the chilling effect on innovation and investment in online services of potentially very high statutory damages awards, *see* DEP'T OF COMMERCE INTERNET POLICY TASKFORCE, WHITE PAPER ON REMIXES, FIRST SALE, AND STATUTORY DAMAGES (2016), at 79-82.

IV. STUDIES 2 AND 3: QUANTITATIVE ANALYSIS OF TAKEDOWN REQUESTS

Studies 2 and 3 present quantitative analyses of individual takedown requests from the Lumen archive. Both studies report on findings from hand coding and machine reading sample requests drawn from a six-month period in 2013. Study 2 analyzes a random sample of over 1,800 takedown requests from the entire 108 million requests archived during the six-month period observed. Study 3 analyzes a separate random sample of over 1,700 requests sent to Google Image Search—a subset of the larger dataset. Each study captures the prominent characteristics of senders, recipients, and targeted works in its sample, and evaluates substantive issues with the notices and underlying claims.

These studies are the first to offer quantitative substantive analysis of claims underlying takedown requests since Urban and Quilter's 2006 work. By and large, the scarcity of research of this kind is due to the difficulty of assembling and coding appropriate data. As discussed more in Study 1, very few service providers, and no major rightsholders, make their notices available for study. The move toward form-based notice submission and public archiving of notices by Google, Twitter, and a handful of other providers has simplified the analysis of some of the broader features of the notice and takedown environment for those providers. Seng's recent work provides an excellent example of how this structured data can be used.[207]

Yet, dealing with the sheer amount of data generated by these automated systems presents a substantial technical challenge. Moreover, form data have some significant limitations. Machine-readable form data provide useful general information, but convey very little information about the judgments made by notice senders and OSP recipients in response to different types of material and varieties of use. Analyzing the substance and validity of the underlying takedown requests requires undertaking a manual examination of requests' subject matter and claims. This is a laborious but necessary process for learning more about sender and OSP practices, the accuracy of claims, and whether OSPs are able to "catch" problematic claims.[208] Developing this knowledge is crucial to understanding how and when notice and takedown succeeds or fails. We sought to fill this gap.

---

[207] Seng, *supra* note 5.

[208] Urban and Quilter took this approach in their 2006 study, for which they coded all of the submissions to the Lumen (then Chilling Effects) archive between January 2002 and August 2005—a total of 876 notices. Urban & Quilter, *Efficient Process*, *supra* note 5, at 641. Now nine years old, the Urban-Quilter study belongs to the pre-industrial era of notice sending, in which the majority of notices were non-standardized, individually sent, and manually reviewed. Broadly, Urban and Quilter found ample reasons to be concerned about notice quality in the DMCA Classic era. Nearly 10% of notices failed to meet the statutory requirements. *Id.* at 674. Another 31% raised serious questions about the underlying copyright claim, including fair use defenses, other substantive defenses, very thin copyright, or non-copyrightable subject matter. *Id.* at 667.

Study 2—covering all of the notices in the six-month set—provides the broadest picture. It confirms the importance of automation and Google's "trusted sender" program for major rightsholders. Indeed, more than 95% of all requests came through Google's Trusted Copyright Removal Program. It also confirms that major senders overwhelmingly request takedown from Google Web Search over the other types of Google services included in Lumen's archive, and that they focus on file sharing and torrent sites. Unfortunately, there is also a significant number of mistaken notices. One in twenty-five requests targeted content that clearly did not match the identified infringed work at all. In total, about a third (31%) raised substantive questions, including problems identifying and locating the disputed works and potential fair use issues.

Study 3 looks only at requests to Google Image Search. Though Image Search received thousands of requests during the period, this represents a tiny subset of all the requests sent to Google. Here, individuals and small businesses were the most frequent senders. Substantive problems were even more apparent in the Image Search sample: seven out of ten (70%) of the Google Image Search requests raised substantive concerns. (Setting aside one individual sender responsible for more than half the requests in this sample (all of which were based on improper subject matter), 36.3% still raised substantive issues).

A. DATA AND METHODS

1. The Lumen Dataset

We requested from Lumen all unredacted takedown requests in its repository for the period between May 1, 2013 and October 31, 2013.[209] This returned 288,675 separate takedown notices. More than one takedown request can be made in an individual notice, and each takedown notice body may contain dozens or hundreds of requests. Accordingly, these 288,675 notices represent well over 100 million (108,331,663) individual takedown requests—i.e., claims of infringement—typically in the form of URLs linking to allegedly infringing material.[210]

By using all of the notices over the six-month period, we were able to build a comprehensive set of Lumen data. At the same time, the dataset from Lumen has some significant limitations for general inquiry into notice and takedown. Most significantly, submission of notices to the Lumen archive is voluntary. Only a few OSPs regularly submit their notices to Lumen, including Google,[211] Twitter, Kickstarter, and a handful of others.[212] This leads to a dataset that covers only a few types of providers. The data is also significantly skewed by the fact that one entity, Google, received 99.4% of the notices in our set. Twitter was the next most prominent recipient, with only .3% of notices. This effect is further pronounced when looking at individual takedown requests rather than notice count: fewer than .008% of

[209] The team at Lumen used the following query to pull the required data from their set: SELECT * from tNoticePriv where date >= DATE_SUB(now(), interval 6 MONTH). E-mail from David Larochelle, Lumen (Feb. 24, 2014, 8:46 PST) (on file with authors).

[210] By way of comparison, Seng's dataset included 539,558 notices submitted to the Lumen archive between January 2001 and December 2012. Seng, *supra* note 5, at 382.

[211] For a list of services for which Google submits notices to Lumen, see *infra* note 214.

[212] Other organizational submitters to Lumen have included at various times Kickstarter, Medium, Proxy.sh, Stack Exchange, Stripe, Tucows, Wikimedia Foundation, and WordPress. E-mail from Adam Holland, Project Coordinator, Lumen (July 29, 2014, 10:10 PST) (on file with authors).

the requests in the dataset were directed to entities other than Google.[213] Further, although Google reports to Lumen on a wide range of its services, it does not include all.[214] Most notably, it excludes takedown requests for its YouTube service.

The dominance of Google notices in our dataset limits our ability to draw broader conclusions about the notice ecosystem. Google has characteristics that set it apart from many other services, including the use of form notices, automated triage systems, and a "trusted sender" program. Google's dominant position in search and the extraordinary number of notices it receives also make it unusual. This makes the Lumen dataset useful for studying an important part of the takedown system,[215] but also means that the characteristics of these notices cannot be extrapolated to the entire world of notice sending.

Our analysis is also limited by the time constraints of the dataset. By using all notices from the six-month period, we were able to balance comprehensiveness with a realistic capacity—the numbers of notices sent to Lumen continue to rise rapidly, and its database now contains over 3 million notices that represent nearly a *billion* takedown requests.[216] However, Study 1 found that notice and takedown activity often fluctuates as senders change targeting activities or techniques, such as by changing the types of content or host sites targeted, or by prioritizing and deprioritizing the sending of takedown notices as an enforcement measure as they implement broader strategies. Notices may come in unpredictable intervals, depending on the practices of the sender. Therefore, our dataset and resulting analysis necessarily reflects a snapshot in time, and may depart from takedown activity trends in other time periods.

A related issue is the time it takes to clean and then code this amount of data. We were attentive to the fact that webpage content may have changed between the time the notice was sent and the time the reviewer examined the request. Many of the problematic characteristics that we identified existed regardless of the dynamic nature of websites. Notices that fail to identify the allegedly infringed work or the allegedly infringing material are key examples of this. In other cases where problematic requests were identified, it was clear that the content on the page was in fact the content that was targeted, but human review revealed the overbroad use of keywords by automated systems or potential fair use defenses.[217] Where certainty was elusive, we chose to err on the side of the notice sender and assumed that missing information would support takedown. Our results overall, therefore, likely undercount flaws. Most notably, in 26% of cases in Study 2, reviewers were unable to

---

[213] We sought to partially remedy the narrowness of this view with our qualitative study, Study 1, which draws from a wide range of service providers, operating across the online ecosystem.

[214] Google contributes notices for the following Google products to Lumen: App Engine, Blogger, Chrome Web Store/Extensions Gallery, Code, Currents, Drive and Docs, G+, Geo – 3D Warehouse, Geo – Panoramio, Google Cloud Storage, Google+ Local, Google Profiles, Groups, Image Search, Orkut, Page Speed Services, Picasa, Sites, and Web Search. E-mail from Shantal Rands Poovala, Google Inc. (May 9, 2014, 11:43 PST) (on file with authors).

[215] In our Study 1 conversations, some major rightsholders made clear that sending to Google Web Search is a significant tactic within their broader strategies. *Supra* Section III.F.

[216] E-mail from Adam Holland, Project Coordinator, Lumen (Aug. 25, 2015, 9:33 PST) (on file with authors). As of August 25, 2015, the Lumen database included 3,164,460 notices that included over 901,320,192 URLs that were the subject of a complaint. *Id.*

[217] For example, a notice sent by BPI requested removal of a song by the artist Usher, but the allegedly infringing material was the movie, "House of Usher." *BPI DMCA (Copyright) Complaint to Google*, CHILLING EFFECTS (Aug. 1, 2013), https://chillingeffects.org/dmca512c/noticecgi?NoticeID=1112904 (Copyright Claim #12, Allegedly Infringing URL #440).

input information about the allegedly infringing material because it had been taken down or the website or webpage on which it resided was no longer live for some other reason.[218] Because we chose to assume that takedown was appropriate in these cases unless there was contrary information, these notices were less likely to be counted as potentially flawed.

Finally, Google also withholds several types of request from Lumen. These exclusions likely resulted in further (and potentially significant) undercounting of flawed notices in our results. Google excludes court orders received by fax, letter, or email. It additionally withholds some, but not all, of the fax, letter, or email notices where content *was not removed* in response to the takedown request.[219] This causes an undercount of notices which, according to Google's evaluation, have substantive flaws. Google also withholds "duplicate requests" where no action was taken on the specified URL because it duplicated a URL specified in a previous request. Google is unable to quantify, and we are unable to estimate, how many duplicate requests are missing from the dataset. This may have caused us to significantly undercount flawed notices, as we had no way of counting requests rejected by Google because they requested takedown of information that had already been removed.

2. Database, Coding Engine, and Sampling Methods

Sampling from and coding a pool of 108 million takedown requests required building a custom database and "coding engine" that allowed us to enter and query inputs about any one takedown request.[220] These tools allowed in-depth investigation of the notices and their component parts by combining available structured data from the form-based submissions with manual coding of characteristics of the sender, target, and claim. We also designed a customized randomization function that supports both sampling across the entire dataset and building randomized "tranches" of more targeted subsets while maintaining overall randomness. This gave us the capacity to focus on narrower categories of notices that may be of great substantive interest but that are unlikely to appear in an overall Study 2 sample because they are "swamped" by more common notices. In this report, the Google Image Search set in Study 3 represents the only tranche we isolated using this method.

We adopted the individual takedown request—rather than the notice body, which might have many requests included within it—as the primary unit of analysis. If a web form allows it, some senders pack many takedown requests, each of which may enforce different copyrights and target different alleged infringements, into one notice body. From the rightsholder's perspective, each request is a claim to remove one allegedly infringing item. From an OSP's perspective, each is a claim that must be acted upon. An accurate analysis thus requires reviewing individual takedown requests. Accordingly, the randomization function presented takedown requests to reviewers in randomized order.

---

[218] In other instances, reviewer could not review AIM in detail because, for example, it was an undownloaded torrent file, because the URL provided linked to more than one potential item on a page that could be the AIM, or because the surrounding context was in foreign language, among other issues.

[219] Google reported that, in total, it received 1,245 non-webform notices during the relevant time period, but was not able to quantify how many of these are missing from the dataset, or how many individual requests these notices contained. Phone call with Fred von Lohmann, Legal Director, Copyright, and Michael Deamer, Legal Assistant, Google Inc., Sept. 4, 2014.

[220] The database is a customized instantiation of OpusData, a MySQL-based subsystem which provides the data and management backend to The Numbers, a movie financial information website.

3. Coding Methods and Data Processing

At the point when the data were imported into the customized database, machine-readable information was extracted and auto-populated into the relevant fields. For example, the coding engine auto-populates the sender's and the recipient's names, the number of links in the notice, whether the sender is a member of the recipient OSP's "trusted sender" program, the name of the allegedly infringed work, if available, and the URL where the allegedly infringing material resides, if provided.

Notices were then coded by the authors and a team of law student research assistants under the supervision of the authors. To ensure consistency, reviewers received training, used a detailed instruction handbook, and were reviewed regularly for accuracy and intercoder reliability. Research assistants and authors met at regular intervals to identify and resolve any arising issues. The authors conducted spot-checks of all reviewers' work.

Data were prepared for statistical analysis in a five-step process. First, the raw data files were exported from the database into CSV files and then into the Stata 13 statistical package;[221] second, sample sizes were confirmed; third, fields from the coding database were matched with the raw data to check for missing data; fourth, standard missing data analysis procedures were employed to properly identify missing data and any potential patterns therein[222] and, where appropriate, replace missing information; and fifth, uni- and bivariate statistical analyses were conducted on the cleaned data to examine frequency distribution and correlative patterns.[223] Please refer to Appendix C for additional information.

B. STUDY 2: IN SIX MONTHS OF LUMEN NOTICES, AN AUTOMATED ONSLAUGHT TO GOOGLE WEB SEARCH

Our first quantitative study considers takedown requests across the entire six-month Lumen dataset. Using the methods described above, we examined the dataset in two ways. Where the source data were machine-readable, we were able to run queries across the entire six-

---

221 For each selection of data there is a unique identifier variable, "link_odid", which distinguishes it from all others. After importation, the sample size of this variable was checked to ensure all of the coded requests properly imported. This confirmed that there were 1,827 general selections and 1,732 image selections.

222 Of primary concern among these selections was to distinguish between legitimately missing data and instances in which the missing data indicated a condition simply did not apply. Each section of questions was examined for skip and missing patterns (to check whether missing data points randomly distributed across the takedown notices and not concentrated among specific types of notices, senders, or reviewers) and then matched to the data in the Stata files. Analysis of skip and missing patterns revealed that the only non-randomly missing information pertained to mutually exclusive question arrays (see below for more details). There were no patterns identified among types of notices, business sectors, or content material. For each section of questions there are two forms of variables: stand alones, whose answers are not dependent or contingent on previous or next questions, and arrays, whose answers are dependent or contingent on previous or next questions. For stand alones, all missing cells remained as such and were not recoded. For example, if the address of the attorney associated with the takedown selection was not included, it remained missing. For arrays, all missing cells were carefully cross-matched with the section of questions they came from. Those that were not applicable rather than missing were recoded appropriately. For example, for each selection there are arrays of questions pertaining to the size of the business that posted the selection. Business size is a mutually exclusive category (a subject cannot be both a small business and a large business). In these cases, selections that were coded as a specific business size were recoded (i.e. the data was relabeled as "0", rather than missing) for all other business sizes. Once these arrays were determined, one set of Stata syntax was written and then edited using find and replace to properly treat each section of the data.

223 Please see Appendix B, Section A.1, for information about obtaining the coded and cleaned data.

month dataset of 108,331,663 requests.[224] For a deeper look, we hand-coded a randomized sample of 1,827 takedown requests,[225] from 1,766 separate notice bodies.[226]

1. Overall Findings: Automation, Major Senders, and Google Web Search Dominate

At the highest level, the Lumen data display two major features: large rightsholders' focus on Google Web Search, and the overwhelming predominance of automated sending by these rightsholders and rights enforcement organizations ("REOs").

- 98.9% of the takedown requests (or 86.7% of the notices) were submitted using an automated Google notice submission form;
- 95.4% of the requests came from members of Google's Trusted Copyright Removal Program ("TCRP")[227]—a program that allows members to submit large volumes of requests.[228]
- 99.8% of the takedown requests in the entire six-month dataset were requests to Google Web Search.

The striking dominance of Google Web Search confirms what we found in Study 1 about the particularized role of search in major rightsholders' use of notice and takedown.[229] Google submits notices for a long list of other services, including social media and cloud hosting services that might be expected to attract notices. Yet compared to Google Web Search, these other services barely register:

---

224 We also used this to cross-check some of our sample results for further assurance that our randomization methodology worked.

225 This number includes only the completed codings and excludes notices that are not DMCA notices and those that contained some other database error (such as where the allegedly infringing material selected for coding is actually a link that the sender provided to the allegedly infringed work).

226 Based on the full dataset of 108,331,663 takedown requests, this sample gives us a margin of error of +/-2.29 at a 95% confidence interval, and +/-3.02 at a 99% confidence interval.

227 Google's TCRP is set up for copyright owners or their enforcement agents ("trusted users") who have a "consistent need to submit thousands of URLs each day" and who have a "proven track record of submitting accurate notices." GOOGLE, HOW GOOGLE FIGHTS PIRACY, *supra* note 193, at 14. *But see* Seng, *supra* note 5, at 417 (finding that the average takedown rate for trusted users was 96.2%, lower than Google's reported overall takedown rate of 97.5% and that for three of the trusted users, the takedown rate was below 86%).

228 This is consistent with numbers released by Google, which reported that at the end of 2012 there were approximately 50 Trusted Copyright Removal Program partners, who together submitted 95% of the URLs submitted during 2012. *See* GOOGLE, HOW GOOGLE FIGHTS PIRACY 14 (2013).

229 *Supra* Section III.F.

TOP 10 GOOGLE SERVICES REPRESENTED IN LUMEN (MAY–OCT 2013)	PERCENTAGE OF REQUESTS IN LUMEN (MAY–OCT 2013)
1. Google—Web Search	99.810%
2. Google—Blogger	.090%
3. Google—Image Search	.030%
4. Google—Docs	.004%
5. Google—Sites	.004%
6. Google—Picasa	.002%
7. Google—Plus (Photos)	.001%
8. Google—Plus	.001%
9. Google—Android Market	.001%
10. Google—Orkut	.001%

Table 2: Top 10 Google Services Represented in Lumen (May–October 2013)

The sheer numbers of notices to Google Web Search and the apparent high level of automation[230] contrast sharply with OSPs' Study 1 descriptions of DMCA Classic processes.

a. Sender Characteristics: Agents Dominate, and a Shift to Movies, Music, and Adult Content Industries

Since Urban and Quilter's 2006 study, there have been two major shifts in the types of senders to Google Web Search. First, the shift towards automation goes hand-in-hand with a shift from major rightsholders sending their own notices to hiring third-party agents to detect infringement and send notices. Second, the major entertainment industries have moved from playing a minor role to become the dominant group requesting Google Web Search takedowns.

i. Third-Party Agents Send the Most Takedown Requests to Google Web Search

Section 512 notices must be sent by either the copyright owner or a person authorized to act on behalf of the copyright owner.[231] Urban and Quilter's 2006 study found that nearly all (98.5%) notices sent to Google services between March 2002 and August 2005 were sent by rightsholders themselves.[232] However, when they reviewed notices sent to a Texas webhost and connectively provider between late 2004 and mid-2007, they saw an emerging role for trade associations and third-party REOs using automated methods to send notices to ISPs.[233] REOs, trade associations, and large owners also began using automated methods

[230] It is theoretically possible that these notices were not generated and sent automatically, but this is highly unlikely given what we learned in Study 1 about the standard industry practices of the types of senders most dominant in the sample.

[231] *See, e.g.*, 17 U.S.C. § 512(c)(3)(A)(vi) (2012) (stating that the notice must include a statement made under penalty of perjury that the complaining party is authorized to act on behalf of the owner of an exclusive right that is allegedly infringed).

[232] Urban and Quilter found that 98.5% of all § 512(d) notices were sent by rightsholders (or their attorneys) directly and that agents, rights enforcement organizations, and trade associations combined sent only 1.3% of § 512(d) notices. Urban & Quilter, *Efficient Process*, *supra* note 5, at 654.

[233] Urban & Quilter, Undue Process, *supra* note 100. Quilter and Heins also identified the emergence of REOs in their 2007 interview study of OSPs. *See* QUILTER & HEINS, *supra* note 5, at 14-17.

to target search. By 2012, Seng found that at least nine out of ten notices sent in 2012 that were archived in the Lumen database were sent by agents rather than by copyright holders themselves.[234] Nearly 46% came from just four large trade associations, led by the UK recording industry association BPI.[235]

Consistent with Seng's findings, the trend towards outsourcing of takedown notices to agents is evident in our sample. Agents of copyright owners sent 91.8% of the requests in our sample; copyright owners themselves sent only 7.5%.[236] *See Fig 1*, below.

SENDER ROLE

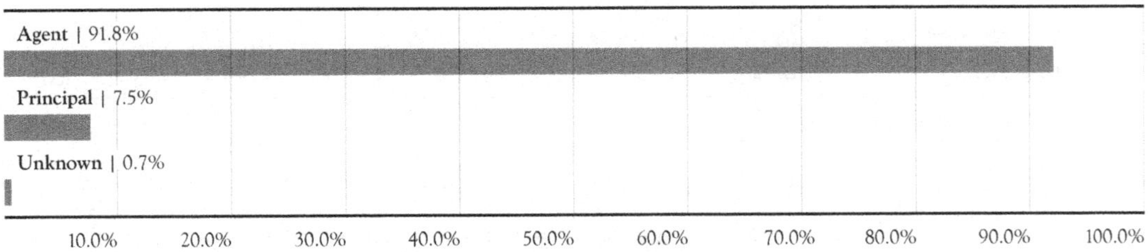

Figure 1: Sender Role

Both third-party REOs and trade associations are highly active senders: REOs sent nearly half (49.3%) of takedown requests and trade associations sent 38%. Law firms trailed far behind at only 0.3%. *See Fig. 2*, below.

AGENT TYPE

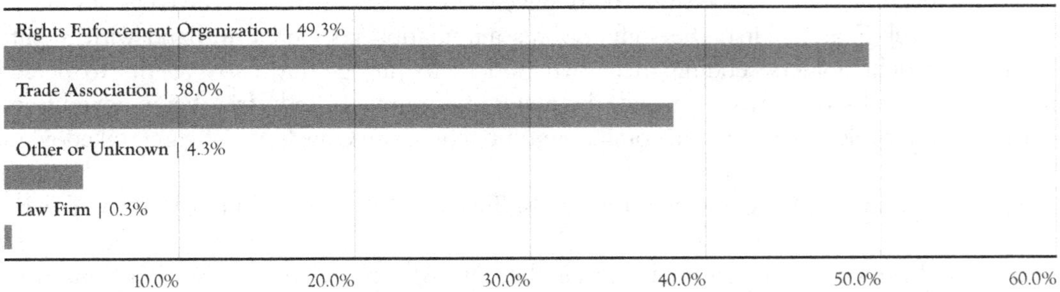

Figure 2: Agent Type

ii. A Major Rise in the Entertainment Industry's Use of Takedown Requests

In their 2006 study, Urban and Quilter found little use of takedown notices to Google Web Search by the entertainment industries; rather the bulk of notices came from small Internet

---

[234] Seng, *supra* note 5, at 448.

[235] In 2012, BPI, IFPI, Publishers Association, and RIAA accounted for 45.7% of all notices sent. *Id.* at 396.

[236] As described in Section III.B.2., individual senders send takedown requests to Google Web Search; however, the relatively low percentage of their contribution to the overall number of requests means that their presence is swamped by the automated requests. It would be useful to isolate and examine requests from individual senders in this set to see if these requests tend to have the same kinds of substantive problems that we identified in the requests sent by individuals and small businesses in Study 3. See *infra* Section IV.C.2.

businesses, computer software companies, and game companies.[237] In that study, the movie and music industries combined were responsible for only 3% of section 512(d) notices.[238]

Today, the story is very different. The music, adult entertainment, and movie/television industries, taken together, now send by far the most significant numbers of takedown requests to Google Web Search. The largest number of requests, 44%, were issued by or on behalf of copyright owners in the music industry, followed by 28.1% requests issued by or on behalf of copyright owners in the adult entertainment industry, and 17% on behalf of copyright owners in the movie/television industry. Other industries make up a significantly smaller proportion of requests: software (7.5%); games (5.4%); books (4.2%); web design (only 0.4%); and photography (only 0.2%).[239] *See Fig. 3* below.

INDUSTRY OF PRINCIPAL

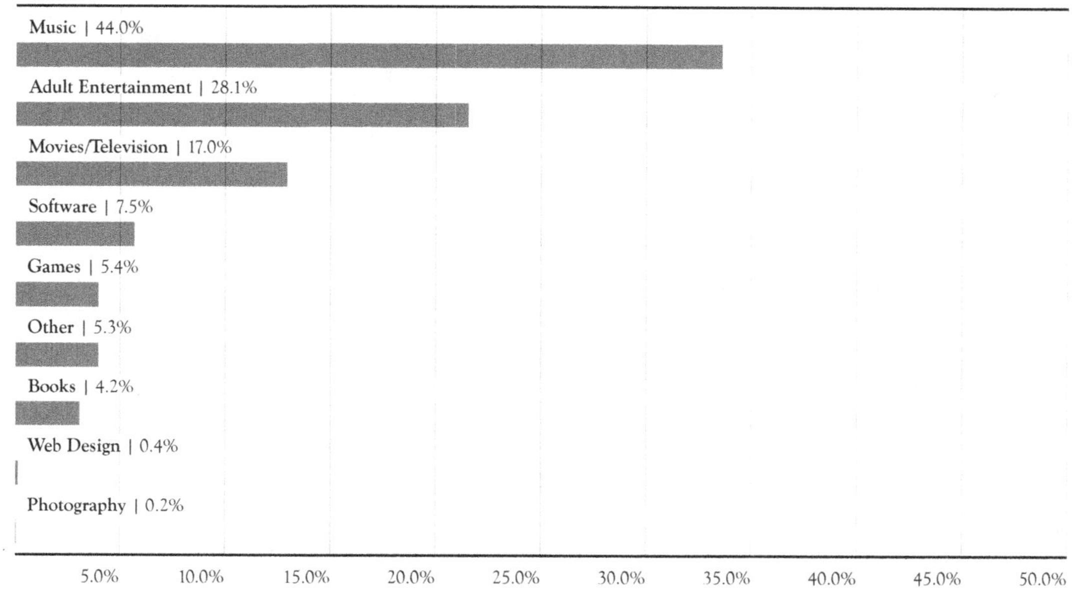

Figure 3: Industry of Principal

When did this shift occur? Seng's work identified a significant shift from the music and movie industry sending practices identified in Urban and Quilter's 2006 study.[240] Seng found that 32.1% of takedown requests sent by the top-fifty notice-senders sent between 2008-2012 were sent by the music industry, 30.1% by the adult entertainment industry, and 20.8%

[237] Urban & Quilter, *Efficient Process*, *supra* note 5, at 651. However, Urban and Quilter found that the movie industry (followed by the computer software and games, and then music, industries) sent the vast majority of § 512(a) notices to ISPs. *Id.*

[238] *Id.*

[239] These numbers add up to more than 100% because copyright owners may be associated with more than one industry.

[240] Seng, *supra* note 5. Urban and Quilter's paper was published in 2006, but their data collection ended in 2005. Urban & Quilter, *Efficient Process*, *supra* note 5. However, in a later conference paper, looking at late 2004 to mid-2007 notices to The Planet, they also identified the emergence of REOs, supporting Seng's finding. Urban & Quilter, Undue Process, *supra* note 100.

by the movie/television industry.[241] Our 2013 data show a similar distribution, though we observed a further uptick in requests sent by the music industry. *See Fig. 4*, below.

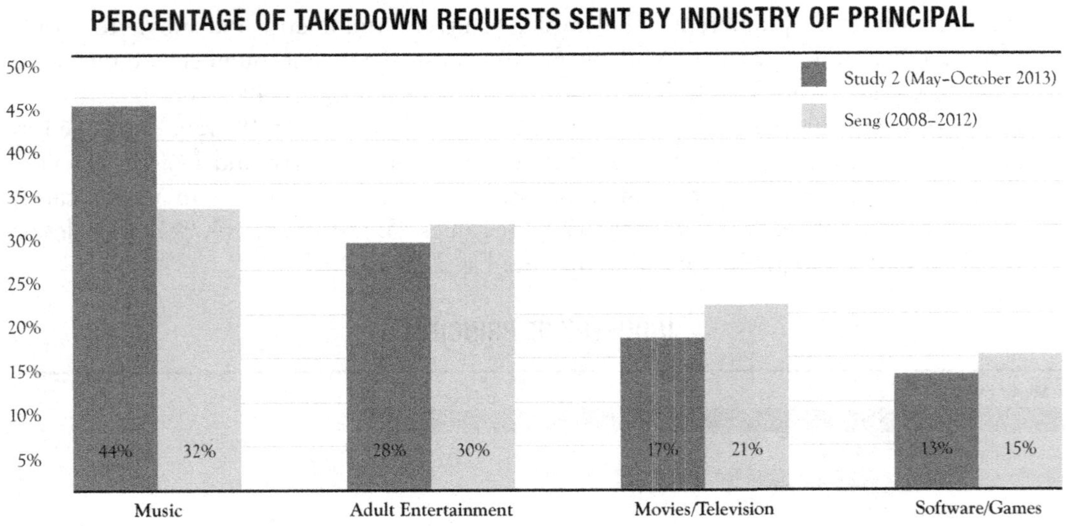

Figure 4: Percentage of Takedown Requests Sent by Industry of Principal

As would be expected from the large representation of entertainment companies in the sample, the types of works for which removal is requested is heavily weighted toward audiovisual works and sound recordings, with pictorial and graphic works making up most of the remainder. *See Appendix A* for details and charts.

b. Target Site Characteristics: Over Two-Thirds of Requests Refer to Torrent or File Search Sites

We also independently classified the type of site ultimately targeted by each takedown request in our sample—that is, the site whose link rightsholders were asking to be removed from Google's Web Search index.[242] In line with what we heard from major rightsholders in Study 1, the large majority—over two-thirds—of the requests in our main sample targeted file sharing sites—predominantly torrent and file search sites.[243] Torrent sites were targeted by 35.9% of requests; and file search sites by 33.9%. A substantial chunk (12.7%) could not be classified. Of the rest, cyberlocker sites[244] (9.9%) made up the most significant group at nearly one in ten, followed by aggregator sites[245] (5.9%), forums/fan sites (4.4%), and video streaming sites (3.7%). A handful were e-commerce sites (0.9%), social media sites (0.5%), or personal websites/blogs (0.3%). *See Fig.* 5, below. Even accounting for the 12.7% we could not classify, a large majority of targets were the type of sites that appear likely to serve infringers. The rather sizable proportion of unclassifiable sites, however, also

[241] Note that Seng's data included notices sent to providers eligible for safe harbors under all of § 512's safe harbors, though, like this main tranche data, the majority of notices were sent to Google Search (and therefore § 512(d) notices). *See* Seng, *supra* note 5, at 419-20.

[242] For Google Web Search, we consider the site that the sender requests be removed from Google's search index to be the "targeted site."For Google Web Search, we consider the site that the sender requests be removed from Google's search index to be the "targeted site."

[243] File search sites are dedicated search sites that link to downloadable content across the Internet.

[244] A cyberlocker site is a site that hosts user files.

[245] Aggregator sites curate and present multiple options for accessing content.

left it unclear whether other types of sites were more common than they appeared from our classifiable numbers.

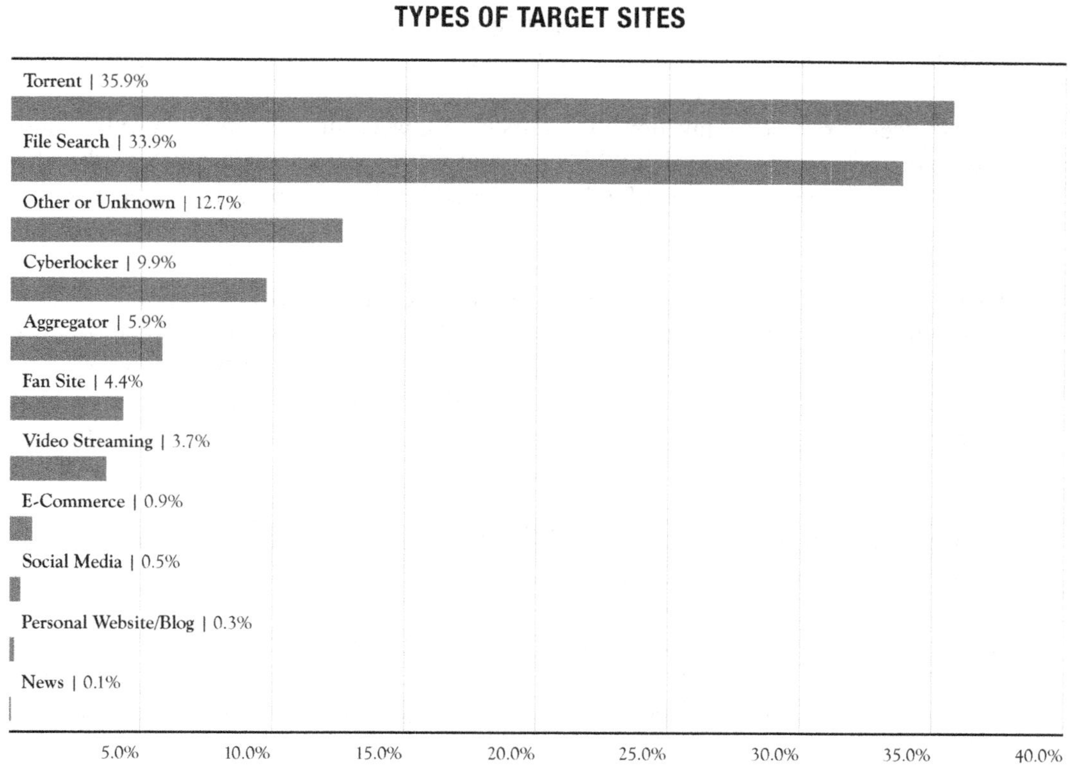

Figure 5: Types of Target Sites

Even with this caveat, the numbers lend credence to major rightsholder claims that they focus on unauthorized file-sharing services when sending notices. Although file search, cyberlocker, and aggregator sites can have legitimate business models, these are also the types of sites most frequently designed and used to share infringing files.

2. Questions of Accuracy and Substantive Judgment

Overall, the general picture that emerged from the Lumen data—an overwhelming focus on Google Web Search, a high level of automation and third-party notice sending, heavy use by major entertainment companies, and a focus on file sharing and torrent sites—still leaves open the question of how accurate these efforts are. As we observed in Study 1, for some senders and for DMCA Auto and DMCA Plus OSPs, notice and takedown has evolved from a low-volume process based on human decision-making to a process dominated by automated systems capable of sending and processing massive numbers of requests. As the scale of the process increases and significant human review becomes impossible, the integrity of the process comes to depend increasingly on the accuracy of these systems. So how accurate are automated notices? To answer this question for our dataset, we examined the substance of each takedown request and its underlying claim of infringement.

Broadly, we found that:

- A few senders—generally targeting unauthorized file-sharing sites—continued to send requests targeting links that led to long-defunct sites, calling into question the checks they do to keep their automated algorithms accurate.

- One in twenty-five of the takedown requests (4.2%) were fundamentally flawed because they targeted content that clearly did not match the identified infringed work. This extrapolates to approximately 4.5 million requests[246] suffering from this problem across the entire six-month dataset.

- Nearly a third of takedown requests (28.4%) had characteristics that raised clear questions about their validity, based solely on the facial review and comparisons we were able to conduct. Some had multiple potential issues.[247] While these requests cannot be described as categorically invalid without further investigation, they suggest that a very substantial number of requests in the six-month dataset—approximately 30.1 million[248]—would benefit from human review.

- 31% of takedown requests were potentially problematic, including both fundamentally flawed requests and "questionable" requests.

- This "questionable" set included requests that raised questions about compliance with the statutory requirements (19.5%), potential fair use defenses (6.6%), and subject matter inappropriate for DMCA takedown (2.3%), along with a small handful of other issues.

These numbers reflect a conservative approach on our part that likely undercounts potential problems. Absent other information, we assumed that notices for material that could no longer be located were valid. We also left out a number of more specific categories. For example, 3.1% of requests referred to sites in languages other than English. We did not count these notices as questionable simply because they targeted non-English language webpages, although these have a high potential of referring to disputes outside of the DMCA's United States jurisdiction and there is no statutory guidance regarding whether OSPs must operate in multiple languages. We also did not count as questionable the 36.8% of Study 2 requests that failed to specify the copyright owner of the allegedly infringed work. Typically, these requests were from trade associations that listed "member companies" as the copyright owner rather than identifying the specific copyright owner of the identified work.

246 The margin of error for our sample is +/-2.29 with 95% confidence, so we can expect a range from 2 million to 7 million in the entire 108.3 million.

247 Several takedown requests had more than one characteristic that put it in the "questionable" category. For the purposes of identifying the total number of questionable claims, we counted these notices only once—in other words, a notice is not counted more than once if it has multiple questionable characteristics.

248 The margin of error for our sample is +/-2.29 with 95% confidence, so we can expect a range from 28.3 million to 33.2 million in the entire 108.3 million.

We delve into each of these categories below.

a. Mistargeting 1: Some Senders Failed to Update Their Algorithms, Continuing to Target Shuttered Sites

In Study 1 interviews, rightsholders identified human cross-checks as a basic standard of good practice for automated noticing. As one of the largest senders described the process, target websites should undergo a human review before they are targeted by any automated search or sending of notices. Sites should then receive periodic follow up reviews to identify changes in behavior. One proxy for how effective this is (or how often it is done) is whether takedown services continue to send notices targeting defunct file-sharing sites.[249] The Google Transparency Report, which breaks down notices by sender and recipient, provides a relatively easy means of tracking such behavior across senders. For example, when Megaupload shut down in January 2012, the Google Transparency Report showed that several major senders, including the RIAA and BPI, responded quickly, dropping out of the pool of notices targeting links to Megaupload. Others, however, continued sending targeting links to Megaupload well into 2014.

We were interested in whether this phenomenon appeared in our dataset. We looked at four major file sharing sites that were shuttered in 2012—Megaupload.com,[250] Btjunkie.org,[251] Filesonic.com,[252] and Demonoid.me.[253] Three of the sites—Megaupload, BTJunkie, and Demonoid—were featured on the MPAA's "notorious markets" list in 2011.[254] The shutdown of Megaupload and arrest of its owners in New Zealand was the signature enforcement action of 2012, resulting in considerable disruption in the file-sharing world. BTjunkie.org and Filesonic.com shut down voluntarily in the wake of the Megaupload seizure.

Our dataset covers May 2013 to October 2013—a period that falls at least eight months and in some cases sixteen months after the closure of the four sites. Most of the large professional senders had stopped targeting these sites well before our inquiry.[255] One however, stood out for continued heavy sending: MarkMonitor/DtecNet. NBC-Universal, Link-Busters, MUSO, Unidam, APCM Mexico, DCMA Force, and Takedown Piracy also continued sending non-negligible numbers of takedown requests to defunct sites. *See Fig.* 6, below.

---

[249] We gratefully borrow this idea from TorrentFreak contributor Enigmax, who anecdotally looked at notices in the Transparency Report targeting some shuttered sites. Enigmax, *Anti-Piracy Outfits Think Megaupload, Demonoid & BT Junkie are Still Alive*, TorrentFreak (Sept. 7, 2012), https://torrentfreak.com/anti-piracy-outfits-think-megaupload-demonoid-btjunkie-are-still-alive-120907/.

[250] See U.S. Department of Justice Press Release, *Justice Department Charges Leaders of Megaupload with Widespread Online Copyright Infringement*, Jan. 19, 2012, https://www.fbi.gov/news/pressrel/press-releases/justice-department-charges-leaders-of-megaupload-with-widespread-online-copyright-infringement.

[251] BTjunkie.org shut down voluntarily in February 2012. *See* Ernesto, *BitTorrent Giant BTJunkie Shuts Down For Good*, TorrentFreak (Feb. 6, 2012), https://torrentfreak.com/btjunkie-shuts-down-for-good-120206/.

[252] Filesonic.com imposed uploading restrictions and disabled file-sharing functionality shortly after MegaUpload was seized, ultimately going offline voluntarily in August 2012. *See* Zach Whittaker, *FileSonic Goes Offline*, CNet (Sept. 3, 2012, 5:12 AM), http://www.cnet.com/news/filesonic-goes-offline/.

[253] Demonoid shut down in August 2012 and returned in January 2014. *See* Ernesto, *Demonoid Down for One Year: The End?*, TorrentFreak (Aug. 3, 2013), https://torrentfreak.com/demonoid-down-for-one-year-the-end-130803/; Ernesto, *Demonoid Returns, BitTorrent Tracker is Now Online*, TorrentFreak (Jan. 9, 2013), https://torrentfreak.com/demonoid-returns-bittorrent-tracker-is-now-online-140109/.

[254] *See* Ernesto, *MPAA Lists "Notorious" Pirate Sites to U.S. Government*, TorrentFreak (Oct. 28, 2011), https://torrentfreak.com/mpaa-lists-notorious-pirate-sites-to-u-s-government-111028/.

[255] *Transparency Report*, Google, *supra* note 12 (see *Specified Domain* subsection).

TOP 8 SENDERS TARGETING FILE SHARING SITES DEAD MORE THAN 8 MONTHS

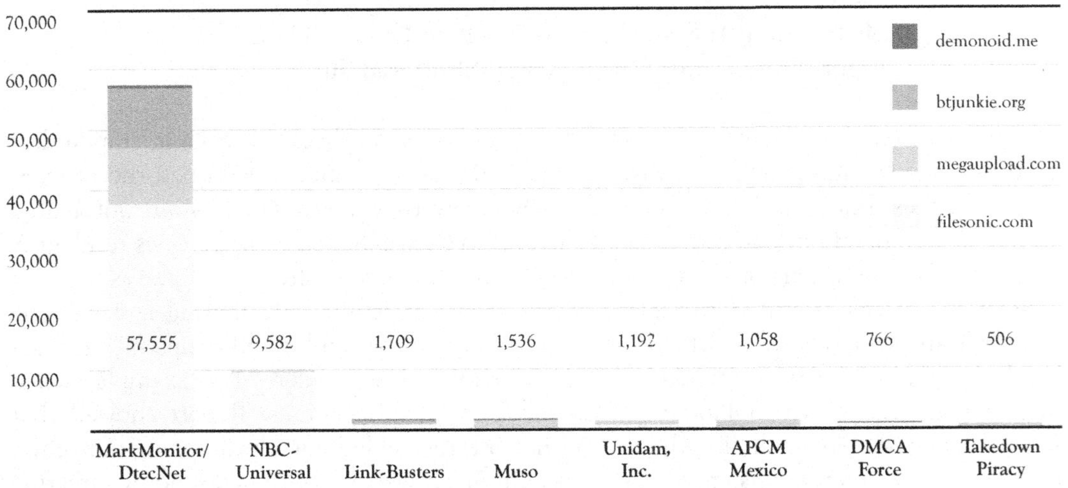

Figure 6: Top 8 Senders Targeting File Sharing Sites Dead More than 8 Months

With the exception of NBC-Universal and APCM Mexico, all of the senders on this list are third-party REOs that act as agents for a wide range of rightsholders. MarkMonitor/Dtecnet is by far the largest entity in the group. It represents Microsoft, Lionsgate, CBS, and Adobe, among many others, and handled notice sending for the Copyright Alert System. Unidam represents the Intellectual Property Promotion Association (an association of Japanese movie studios) DMM.com Labo (a distributor of Japanese adult videos), among others. Link-Busters represents primarily small record labels and some larger firms like Random House.

These findings suggest that more human judgment and review are needed at some agent services. In our Study 1 interviews, some large rightsholders described employing a mix of REOs, shopping—as one respondent described it—in a competitive market differentiated by strategies, technologies, and promises of "silver bullets." One of the challenges of this process, the respondent noted, was the need to continually monitor the efficacy and quality of contracted services. In an environment in which there is little or no cost for mistaken takedowns, there is correspondingly little pressure on the senders to provide safeguards. As we discuss in Section V below, our results suggest that rightsholders should capitalize on competition in the REO market by demanding that services differentiate themselves based on a minimum set of standards for accuracy and human checks.

b. Mistargeting 2: Targeted Material Does Not Match the Allegedly Infringed Work

In about one out of twenty-five (4.2%) takedown requests, the allegedly infringed work ("AIW") described in the request did not match the allegedly infringing material ("AIM") at all. (We use these abbreviations frequently in this part of the study.) On closer examination, these tended to fall into a few categories.

Sometimes the AIM was connected to the principal identified in the notice, but clearly was not a match to the AIW. For example:

- A request[256] sent by REO Vobile on behalf of Paramount where the AIW was the Paramount movie "An Officer and a Gentleman" and the AIM was the Paramount movie "Anchorman: The Legend of Ron Burgundy."

- A request[257] sent by REO Attributor on behalf of Cambridge University Press where the AIW was the book "Cultural Politics in Human Rights: Comparing the US and UK" and the AIM was another book published by Cambridge University Press, "The Unfinished Peace After World War I."

In these cases, it is difficult to see how the mismatch occurred, as the titles are so different. Other times, the mismatch was apparently a result of automated systems making over-broad uses of key words when searching for allegedly infringing material and generating notices. This happened especially when the key words were not particularly unique phrases or words. For example:

- A request[258] sent by BPI where the AIW was the song "Change of Heart" by the artist Change, and the AIM was a torrent file for an episode of the television show "How I Met Your Mother" titled "A Change of Heart."

- A request[259] sent by BPI where the AIW was copyrighted works by artist "Usher" and the AIM was a file of the movie "The House of Usher."

- A request[260] sent by REO IP-Echelon on behalf of HBO where the AIW was the television series "Girls" and the AIM was an episode of the television show "New Girl."

- A request[261] sent by BPI where the AIW was the song "Free" by the band The Bees and the AIM was the documentary "Vanishing of the Bees."

Other clear mismatches occurred when the provided URL led to a search results page that did not actually list content related to the AIW. Anecdotally, we observed that the further down the list of search results the targeted page, the less likely the material on that page was to match the AIM. This could be because the dynamic nature of search results amplifies differences as one moves further and further down the list of results. In any case, it reveals a flaw in pointing to pages of results, rather than specific, problematic links. For example:

---

[256] *DMCA (Copyright) Complaint to Google*, CHILLING EFFECTS (Aug. 7, 2013), https://chillingeffects.org/dmca512c/noticecgi?NoticeID=1123159 (Copyright Claim #3, Allegedly Infringing URL #40).

[257] *DMCA (Copyright) Complaint to Google*, CHILLING EFFECTS (June 2, 2013), https://chillingeffects.org/dmca512c/noticecgi?NoticeID=1001650 (Copyright Claim #7, Allegedly Infringing URL #3).

[258] *BPI DMCA (Copyright) Complaint to Google*, CHILLING EFFECTS (May 16, 2013), https://chillingeffects.org/dmca512c/noticecgi?NoticeID=973630 (Copyright Claim #22, Allegedly Infringing URL #150).

[259] *BPI DMCA (Copyright) Complaint to Google*, CHILLING EFFECTS (Aug. 1, 2013), https://chillingeffects.org/dmca512c/noticecgi?NoticeID=1112904 (Copyright Claim #12, Allegedly Infringing URL #440).

[260] *DMCA (Copyright) Complaint to Google*, CHILLING EFFECTS (Oct. 1, 2013), https://chillingeffects.org/dmca512c/noticecgi?NoticeID=1233091 (Copyright Claim #1, Allegedly Infringing URL #20).

[261] *BPI DMCA (Copyright) Complaint to Google*, CHILLING EFFECTS (Oct. 7, 2013), https://chillingeffects.org/dmca512c/noticecgi?NoticeID=1246501 (Copyright Claim #185, Allegedly Infringing URL #5).

- A request[262] sent by REO MarkMonitor on behalf of Microsoft where the AIW was a list of Microsoft software products and the search results page identified in the request did not include any Microsoft content. Instead, it included other material that matched some of the search terms ("Microsoft home and student serial"), such as an episode of the television show "Lost" called "There's No Place Like Home" and a television show called "Extreme Makeover: Home Edition." In this case, the search results page identified in the request was the 397th page of results.

- A request[263] sent by REO IP-Echelon on behalf of HBO where the AIW was the talk show "Real Time with Bill Maher" and the search results page identified in the notice was the 280th page of results for the search term "real estate agent." This search results page did not return any related content at all.

In the worst mismatches, the AIM appeared to be entirely unrelated to the AIW, the rightsholder, or the key words or search terms used to generate the takedown request:

- A request[264] sent by REO Morganelli Group on behalf of Lionsgate where the AIW was the movie "Haunting in Connecticut 2: Ghost of Georgia" and the AIM was torrent file of a television episode in the "Army Wives" series, which is a Lifetime series (owned by the Hearst Corporation and Disney).

- A request[265] sent by REO IP-Echelon on behalf of HBO where the AIW was the television series "Newsroom" and the AIM was a Dutch-language e-book.

c. Meeting and Not Meeting the DMCA's Statutory Due Process Requirements

The DMCA imposes a number of requirements on takedown notices aimed at ensuring basic due process and preventing abuse. These include statements that the owner or an authorized agent is the one requesting removal; that the claim is accurate and in good faith; that there is information sufficient for a target to respond to the claim; and more substantively, that there is sufficient information about the AIW and AIM to allow the OSP to locate the AIM and assess the request.[266]

A surprising number of takedown requests—19.5%, a little less than one in five—raised questions about whether the request was in compliance with these basic elements of notification.

---

262 *DTecNet DMCA (Copyright) Complaint to Google*, CHILLING EFFECTS (July 9, 2013), https://chillingeffects.org/dmca512c/noticecgi?NoticeID=1068330 (Copyright Claim #2, Allegedly Infringing URL #714).

263 *DMCA (Copyright) Complaint to Google*, CHILLING EFFECTS (May 1, 2013), https://chillingeffects.org/dmca512c/noticecgi?NoticeID=947424 (Copyright Claim #6, Allegedly Infringing URL #58).

264 *DMCA (Copyright) Complaint to Google*, CHILLING EFFECTS (July 4, 2013), https://chillingeffects.org/dmca512c/noticecgi?NoticeID=1060620 (Copyright Claim #7, Allegedly Infringing URL #71).

265 *DMCA (Copyright) Complaint to Google*, CHILLING EFFECTS (June 14, 2013), https://chillingeffects.org/dmca512c/noticecgi?NoticeID=1023535 (Copyright Claim #20, Allegedly Infringing URL #37).

266 17 U.S.C. § 512(c)(3)(A)(i)-(vi) (2012).

i. Notice Webforms Appear to Minimize Problems with Technical Statutory Requirements

We saw only a few issues with the more technical of the statutory requirements: a signature of an authorized party;[267] information reasonably sufficient to contact the complaining party;[268] a statement of good faith;[269] a statement that the notice is accurate;[270] and a statement under penalty of perjury that the complainant is authorized to act on behalf of the copyright owner.[271] Indeed, only a handful of requests in our sample raised questions about these requirements.

We expect this is in no small part due the professionalized nature of the senders in our sample, some rightsholders' use of the standardized Automated Copyright Notice System, and Google's shift to a webform that demands these pieces of information. As noted above, users of Google's TCRP sent the vast majority of these notices—all pre-vetted, knowledgeable senders that make regular use of the takedown system. As regards the webform, all of the requests in the coded set were submitted using Google's online form. In Study 1 interviews, OSPs suggested that using webforms increase the likelihood that the required technical statements are included. For example, Google's webform system will not accept a takedown request unless senders provide an email address where they can be contacted.

ii. Based on a Conservative Metric, a Significant Number of Requests Failed to Adequately Identify the Works at Issue

Deficiencies in the remaining two statutory requirements—to sufficiently identify both the allegedly infringed work[272] and the allegedly infringing material[273]—were far more common. Nearly all of the notices that raised statutory compliance issues presented one of these two problems. These are two of the most substantively important statutory requirements. Without the ability to compare the allegedly infringed work with the material for which takedown is requested, an OSP (or the responsible user, or another reviewer) cannot assess whether a takedown request is proper. Accordingly, notices that do not substantially comply with these requirements are not sufficient to confer knowledge of infringement—and thus potential secondary liability if the OSP does not remove the material—on the service provider.[274] Unsurprisingly, these are some of the most litigated of section 512's requirements. Rightsholders and OSPs have regularly tussled over allocating the costs of identifying and removing infringements.[275]

The largest problem group, making up 13.3% of the takedown requests in our sample, made it difficult to locate the allegedly infringing material targeted for removal. Most often, the

---

267 § 512(c)(3)(A)(i).

268 § 512(c)(3)(A)(iv).

269 § 512(c)(3)(A)(v).

270 § 512(c)(3)(A)(vi).

271 *Id.*

272 § 512(c)(3)(A)(ii).

273 § 512(c)(3)(A)(iii). This element also requires that senders provide information reasonably sufficient to permit the service provider to locate the material. *Id.*

274 § 512(c)(3)(B)(ii).

275 *See, e.g.*, *infra* note 277.

In 13.3% of requests, it was difficult to locate the allegedly infringing material—typically because the request provided a URL that led to a search results page or aggregator page that included multiple works.

notice provided a URL that led to a search results page or aggregator page that included multiple works, making identifying the AIM problematic. Google's introduction of webforms appears to have made no positive difference here. Issues with identification were even more frequent than they were in Urban and Quilter's 2006 study.[276]

Problems with the request's identification of the allegedly infringed work arose in at least 6% of takedown requests.

Problems identifying the allegedly infringed work also arose in a significant number—at least 6%, or about one in sixteen—of takedown requests. This is one of the most contested issues in debates over section 512, especially the question of when a sender's "representative list" of allegedly infringed works is sufficient.[277] For the purposes of this study, we took a conservative approach to flagging requests as questionable for failing to identify the AIW; 6% is probably an undercount of the notices that would raise this question under the current case law. Requests were counted as questionable only when the sender failed to identify any specific work at all: when the only information was a link to a website homepage, or when it was a link to another webpage where an AIW was not apparent. If the sender provided *any* additional information, such as a title for the work or a general

[276] Urban & Quilter, *Efficient Process*, *supra* note 5, at 674 (finding that 1 in 11 notices displayed significant statutory flaws). Urban and Quilter counted these notices slightly differently—they included notices that did not specify the complainant's contact information—but this difference only increases the relative proportion of notices with identification problems in our sample. *See id.*

[277] Where multiple copyrighted works at a single site are covered by the same notification, section 512 permits a sender to identify the allegedly infringed copyrighted work through a representative list of such works. 17 U.S.C. § 512(c)(3)(A)(ii). In *Perfect 10 v. Google*, the court held that a reference to the totality of the sender's image collection does not identify what may have been infringed and is not a representative list under the statute. Perfect 10, Inc. v. Google, Inc., No. CV 04-9484 AHM SHX, 2010 WL 9479060 (C.D. Cal. July 30, 2010), *aff'd*, 653 F.3d 976 (9th Cir. 2011). Therefore, the notices sent that identified the copyrighted works by referencing an electronic folder of 15,000 images and offering the service provider a username and password to access the sender's website was not sufficient to identify the copyrighted work as required by the DMCA. *Id.* at *9. A list of artists' names without specifying any particular songs or allegedly infringing links is also insufficient to constitute a "representative list." Arista Records, Inc. v. Mp3Board, Inc., No. 00 CIV. 4660 (SHS), 2002 WL 1997918, at *15 (S.D.N.Y. Aug. 29, 2002) ("Although the DMCA permits a copyright owner to identify a 'representative' list of works... in this case, a bare list of musical artists whose songs were allegedly linked to did not constitute a representative list of works, or notice equivalent to a list of representative works that can be easily identified by the service provider."). On the other hand, providing a web address to where copyrighted works are located in conjunction with other identifying information may satisfy the requirements of providing a "representative list" as well as the requirement to provide information reasonably sufficient for the service provider to locate the infringing material. ALS Scan v. RemarQ Cmtys., Inc., 239 F.3d 619, 625 (4th Cir. 2001) (finding that the sender provided a representative list of infringing material and was sufficient to enable the defendant to located the infringing material where the sender 1) identified two of the defendant's newsgroups that were created for the sole purpose of publishing ALS Scan's copyrighted works, 2) asserted that virtually all of the images on the two sites were plaintiff's copyrighted material, 3) referred the defendant to two web addresses where it could find photographs of its models and obtain its copyright information, and 4) noted that its material could be identified because it included plaintiff's name and/or copyright symbol.). In *Perfect 10, Inc. v. Giganews, Inc.*, the district court found that Perfect 10 had neglected the obvious means of identifying material (Message-ID) in favor of screenshots of search results, which it viewed as not "reasonably sufficient to permit the service provider to locate the material." Perfect 10, Inc. v. Giganews, Inc., No. CV 11-07098-AB SHX, 2014 WL 8628031, at *8 (C.D. Cal. Nov. 14, 2014) (quoting 17 U.S.C. § 512(c)(3)(A)(iii)), *aff'd*, No. 15-56026, 2017 U.S. App. LEXIS 1128 (9th Cir. Jan. 23, 2017).

description of it, then we assumed that the information was sufficient to identify the AIW and did not count the notice as questionable. This likely excluded a large number of notices that could be considered questionable using a less conservative metric. For example, many notices from the adult entertainment industry identify the AIW by including a very vague description such as "video and image series by [principal name]" and a link to a website with a list of hundreds or even thousands of titles or images. If any such description was included, no matter how vague, we did not count that request as problematic.

The contrast between the very low incidence of technical statutory problems and the much higher instance of identification problems is striking. Overall, web forms appear to preserve compliance with the more technical statutory requirements, but do not increase compliance with the more substantive identification issues. This is relevant to the allocation of responsibility and the availability of the safe harbor for OSPs. As noted, insufficiently identifying the copyrighted work or the allegedly infringing work can render a notice ineffective in conferring knowledge on OSPs. In contrast, notices that have the more technical statutory deficiencies—such as a failure to substantially comply with the requirements for a signature, good faith statement, and statement that the notice is accurate and that the complaining party is authorized—may still be considered in determining whether and OSP has actual or red flag knowledge of infringement.[278]

d. One in Fifteen Notices Presented a Fair Use Question

From its inception, the notice and takedown regime has prompted concerns about whether its procedural structure sufficiently supports fair use and other copyright limitations. In a copyright lawsuit, a defendant has the ability to raise fair use or other issues in defense; in notice and takedown, she has the theoretical ability to file a counter notice. Yet Study 1 and other research on this issue consistently shows that counter notices are rarely used.[279] Questions about whether the counter notice process is sufficient to protect expression have increased with automation—fair use analysis is famously fact-specific and nuanced, and generally considered ill-suited for automated decision-making. How to avoid catching fair use "dolphins" in automated nets set to catch infringements has become an even more pointed issue since the Ninth Circuit Court of Appeals, in *Lenz v. Universal Music Corporation*,[280] held that copyright owners must consider fair use before sending a notice. Here, we consider whether fair use presents a substantive issue in our heavily automated dataset.

About 1 in 15 (6.6%) of requests were flagged with characteristics that weigh favorably toward fair use.

About one in fifteen (6.6%) requests had at least one characteristic that likely weighs favorably toward fair use. These requests predominantly targeted such potential fair uses as mashups or remixes, or links to search results pages including mashups or remixes. An additional notable group of requests targeted ringtones. The remainder varied widely—from cases where the AIM copied only a small portion of the AIW to cases where,

[278] 17 U.S.C. § 512(c)(3)(B)(ii).

[279] Quilter & Heins, *supra* note 5; Seng, *supra* note 5; Urban & Quilter, *Efficient Process*, *supra* note 5.

[280] Lenz v. Universal Music Corp., 2016 U.S. App. LEXIS 5025, at *16 (9th Cir. Mar. 17, 2016) (holding that fair use is "authorized by law" and a copyright holder must consider the existence of fair use before sending a takedown notification under § 512(c)).

based on surrounding information, the AIM was apparently being used for educational or instructional purposes.[281]

We could not do a full fair use analysis, and focused on characteristics that reviewers could observe and record relatively easily. The final merit of any potential fair use claims within this set will vary. Our goal was to observe whether automated systems appeared to generate any significant number of notices for which more contextualized human review is needed to check for fair use. It appears that they do: around 7 million notices out of the full 108.3 million can be expected to present these issues.[282]

In Study 1, some rightsholders described a variety of tactics they use to avoid targeting potential fair uses. We pick up some of these ideas, and add others, in our recommendations.

e. Subject Matter Other than Copyright

A smaller number of requests raised questions by appearing to use the DMCA's takedown measures—which apply only to copyright infringement—when other concerns were involved. Trademark concerns were explicitly or implicitly raised in 1.3% requests. Anti-circumvention issues—which typically arise where the AIM is a product key, rather than the copyrighted work it protects—were explicitly or implicitly raised in 1.1%.

Because these numbers are small enough to be within our margin of error for the Study 2 sample—+/- 2.29% at a 95% confidence level—we do not draw conclusions. The Google Image Search takedown requests we evaluate below in Study 3, however, present many more such issues.

f. Other Issues

Other issues arose in smaller numbers. These included requests that incorrectly identified an REO as the copyright owner; those that used very vague identifying terms for the copyright owner (such as "self"); requests targeting material that was actually embedded on the page linked to in the notice (and thus hosted elsewhere);[283] requests targeting pages linking to content that appeared to be authorized by the copyright holder; and one instance where the sender was apparently targeting a competitor.

3. Study 2: Discussion

To a large degree, analysis of our full six-month dataset bears out the main themes we heard in Study 1 from DMCA Plus OSPs and rightsholders: a high degree of automation directed at some OSPs, a focus on protecting major entertainment industry copyrights, and a focus on targeting the most obvious infringement sites.

It also suggests some problems. Despite the efforts of some major senders to limit algorithmic mistakes, nearly a third of requests presented serious questions about their validity. Potential

---

[281] We mentioned "covers" in the original version of this paper, which caused some confusion. Being a cover was *not*, in itself, a characteristic that we counted as weighing towards fair use. We have removed the reference and regret the confusion. See Appendix C for a further explanation of our methodology.

[282] The margin of error of for our sample is +/-2.29 with 95% confidence, so the expected range of requests in this category is between 4.7 million and 9.6 million.

[283] For example, a YouTube video embedded on a blog. As this is analogous to hyperlinking to content, liability for copyright infringement in this situation is questionable. See *supra* note 184.

fair use defenses and outright mistargeting of materials that do not match the AIW are most concerning, followed closely by substantive problems with identifying the AIW and AIM. Some categories of potential issues are proportionally small—4% or 6%—but the sheer number of notices means that even these single-digit percentages each translate into millions of potential problems across the entire six-month set. *See Fig. 7*, below. While it is tempting to focus on the relatively small percentages, each mistake potentially affects an individual target's expression—millions of mistakes are concerning, regardless of whether many millions more hit the mark.

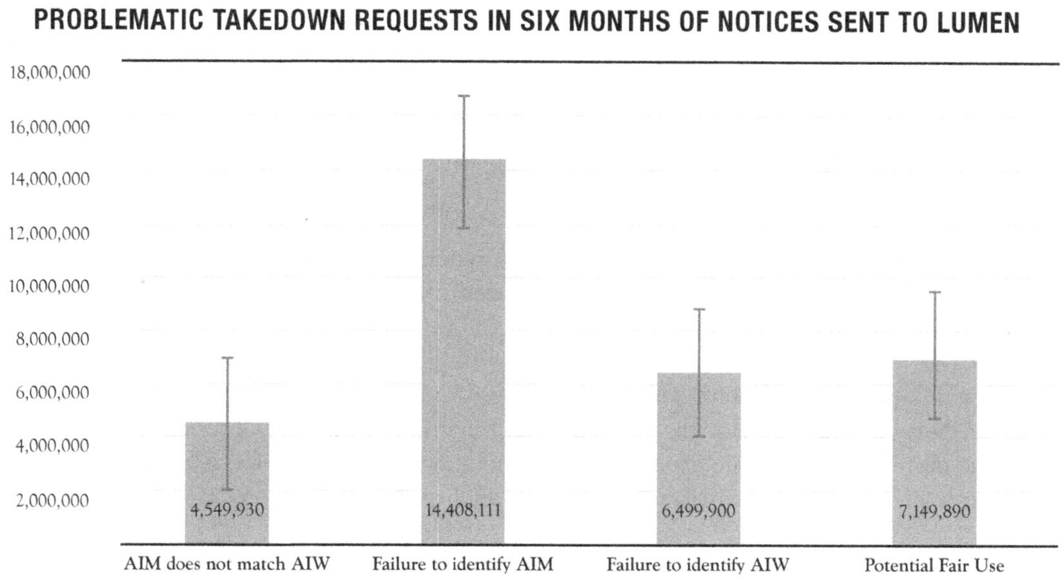

Figure 7: Problematic Takedown Requests in Six Months of Notices Sent to Lumen (Showing Error Bars Based on Margin of Error of +/-2.29 at 95% Confidence Interval)

This state of affairs suggests that the notice and takedown process, as practiced in our cohort of notices, imposes a high burden on those mistaken targets. Where there is uncertainty, OSPs in Study 1 generally described taking conservative approaches that favored takedown. Notices to Google Web Search—the vast majority of the notices here—are even less likely to prompt counter notice and putback than notices to hosting providers: the DMCA imposes no duty on search providers to notify targets, which have no service relationship with them.

In Study 1, rightsholders stressed the importance of automation to address large-scale infringement. They described their approach as targeting the "worst of the worst" infringing sites, and described employing various checks to avoid mistakes. Some of the questionable notices in our sample do likely reflect relatively minor sins of misidentification that are unlikely to raise significant worry about freedom of expression. Still, the numbers are too high to dismiss as an acceptable cost of enforcement. Automation allows rightsholders to practice large-scale takedown, but algorithmic mistakes magnified and multiplied by the millions suggest that it should be done better, and that more human review and crosschecks are required.

In Section V, below, we offer some recommendations for practices that may help manage some of the issues with automation, drawn from some of the best practices rightsholders and OSPs described in Study 1. We also suggest some reforms that would give targets better ways to address problematic notices.

C. STUDY 3: IN SIX MONTHS OF NOTICES TO GOOGLE IMAGE SEARCH, SMALLER COPYRIGHT HOLDERS AND MANY MISTAKES

Our second quantitative study looks only at DMCA notices sent to Google's Image Search service.[284] Using the methods described above, we defined a subset, or "tranche," of data from our full six-month dataset that included all, and only, notices sent to Google Image Search. The Google Image Search tranche contained 33,409 requests,[285] housed in 2,777 notices, to remove links from the Google Image Search index. We then directed the randomization engine to provide a randomized sample of that subset. We reviewed and coded a randomized sample of 1,732 of these Google Image Search takedown requests[286] from 607 unique notices.

We chose Google Image Search for our tranche because we expected it would provide a different profile from the Google Web Search notices that dominate the full six-month set. For example, we expected to find more photographers and visual artists in the pool of notice senders.

1. Overall Findings: Small Copyright Owners Acting for Themselves, Significant Problems, and the Outsized Effect of One Determined Sender

The Google Image Search results were indeed strikingly different. First, the dataset is far smaller—Image Search links are targeted much less frequently than Web Search links. While the 33,409 requests in the Google Image Search tranche is a substantial number, it is far less than the more than 108 million requests, overwhelmingly targeted to Google Web Search, in the full six-month set from which we drew the tranche.

Second, the entertainment industries' dominant presence in the Study 2 requests—which were nearly all sent to Google Web Search—is barely to be found in Google Image Search, with only the adult entertainment industry significantly represented. Instead, the vast majority of takedown requests came directly from individuals and small businesses, requesting takedown for a mix of reasons. Relatively smaller chunks of notices came from

---

[284] These requests all relate to links to allegedly infringing material indexed by Google for its Image Search service; accordingly, like Web Search, the relevant safe harbor protections fall under sections 512(d) (location services) and 512(b) (caching). 17 U.S.C. § 512(b), (d) (2012).

[285] This number is an approximation because of the manner in which structured data from Google's Image Search complaint web form interacts with our coding engine algorithm. For Image Search, Google requests three separate URLs for one instance of infringement: a link to search results page, a hotlink to hosted image file, and a link to the page where the file appears. The data structure does not allow us to distinguish between this situation and a notice that contains multiple separate takedown requests URLs and unfortunately, senders often did not comply with the three-link request, leaving us without a simple calculation to estimate unique requests in the entire Google Image Search set. We did, however, record when a sender in our sample provided more than one URL to identify the alleged infringement. Only 15.2% of the non-Miller requests provided more than one URL to identify the alleged infringement (with 7% providing all three URLs requested by Google). We note that nearly two-thirds of these are from just two senders. Finally, the full count of 33,409 requests includes a number of requests that our coding engine algorithm improperly identified as AIMs. These incorrectly selected requests were excluded from our sample and analysis.

Accordingly, the margin of error for this sample—+/-2.29 at a 95% confidence interval—is artificially inflated because it is based on the full count of 33,409 requests. Because some links are irrelevant repeats of the same request in different form, this is a conservatively wide margin of error.

[286] As with the sample from the main population, this number includes only DMCA requests. In addition to the exclusions discussed above, *supra* note 225, approximately 40 requests were excluded from the set because they possibly referred to child pornography. We instructed reviewers to immediately cease coding any such requests, did not review them, and did not include them in this analysis.

professionalized adult entertainment providers, visual artists and designers, and photography providers. And unlike the Web Search Study 2 sample, the Image Search tranche contained very few agent senders. These differences are perhaps unsurprising, given the movie and music industry's limited focus on still images. They indicate how important it is to explore the use of notice and takedown across a variety of types of services.[287]

Third, as described below, the Google Image Search data provide a striking example of the ability of a small number of determined senders to "flood" the recipient OSP with takedown requests. Indeed, *one* sender—an individual, not a professionalized sender—sent more than half the requests in the sample.

Seven out of ten (70%) of the Google Image Search requests raised substantive questions about the claim the takedown request was based on.

Finally, Google Image Search requests were much more likely to raise a substantive concern: seven out of ten (70%) of the Google Image Search requests raised substantive questions about the claim the takedown request was based on. Leaving aside the one sender responsible for more than half the notices (all of were based on improper subject matter), 36.3% still raised substantive issues. As in the Study 2 sample, this "questionable" set includes requests that were potentially flawed based on the subject matter of the request and those that raised potential fair use defenses. In addition, this set includes requests that exhibited potential ownership issues, failed to identify the allegedly infringing material, or targeted material that appeared likely to be in the public domain.

At the same time, Google Image Search requests were much *less* likely to present issues with identifying the works in question; this could be due to the apparently much smaller role of automation in these notices.

a. Sender Characteristics: Very Few Agents, Small Copyright Owners Acting for Themselves, Extra-Territorial Disputes, and One Highly Prolific Sender

The mix of notice senders to Google Image Search looks very different from those sending requests to the more general-purpose Google Web Search. Whereas agent senders dominate the Web-Search-heavy overall dataset, principals—that is, copyright claimants themselves, rather than agents acting for them—sent the vast majority of takedown requests in the Google Image Search tranche. Most of these senders appeared to be individuals with no classifiable industry, though the adult entertainment industry, visual artists and designers, and photographers made notable showings. Members of Google's Trusted Copyright Removal Program, composed of many large, professional senders, did not send *any* of the requests to Google Image Search, though several senders submitted significant numbers of requests.

A striking feature of the dataset is the role of one individual, whom we will call Ella Miller.[288] Miller is a European individual who is embroiled in an online dispute about modeling photographs taken of her—the precise character of which was not evident to us. Apparently in response to discussion of these photographs on various web forums, Miller sent thousands

[287] We note that these findings also cannot necessarily be extended beyond Google Image Search to other image services—for example, Flickr, Photobucket, or Instagram, which, as hosting services known for images, may attract more professional attention.

[288] This is a pseudonym.

of takedown requests to Google Image Search, representing over half (52.9%) of the requests in the sample. Most pointed to written material she alleges is defamatory, harassing, slanderous, or threatening.

Because of Miller's outsized role and her general failure to send notices that are proper subject matter for takedown, we have frequently separated out the non-Miller notices for analysis. Miller's notices are as much a feature of the set as any other sender's notices, but it can also be helpful to look separately at her notices because some of their features would otherwise obscure observations that are more typical across all senders. For example, Miller nearly always complains of links to written work, while others, as would be expected of notices to an image search service, nearly always complain of links to visual works.

i. Principals, Not Agents

Copyright owners themselves—principals—sent 94.4% of the requests to Google Image Search—a major difference from the agent-heavy Google Web Search requests.[289] Agents sent only a handful (.5%).[290] The remaining 5.1% could not be classified. Setting aside Miller's requests, principals sent nearly nine out of ten (88.1%) of the requests, agents sent 1.1%, and 10.8% could not be classified. *See Fig.* 8, below.

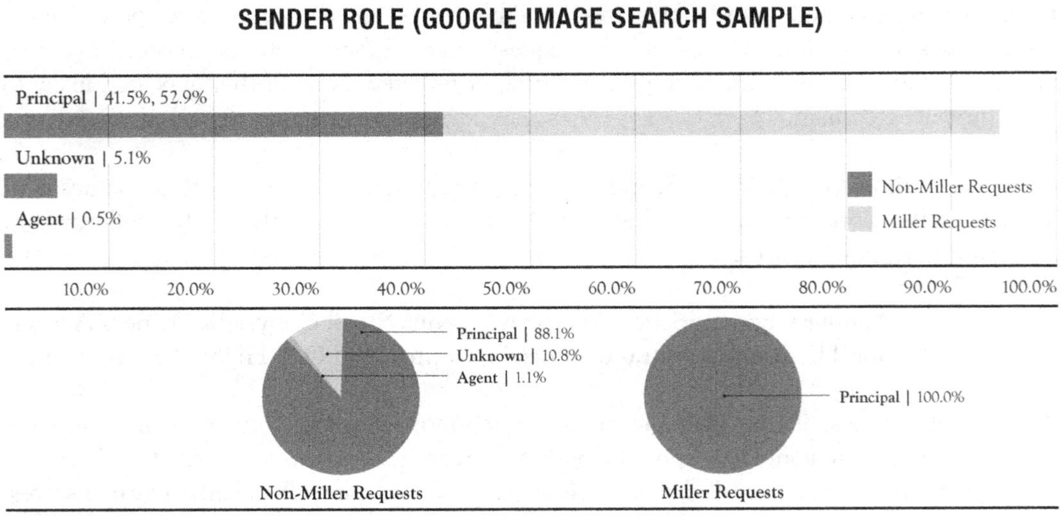

Figure 8: Sender Role (Google Image Search Sample)

ii. Principal Size: Individuals and Small Businesses

Also unlike Study 2, the large majority of Google Image Search senders were individuals or small businesses.[291] Small businesses with fewer than 100 employees—and often many fewer—sent 41.1% of the non-Miller requests. Only a handful of senders were "small-

[289] Principals were categorized as such where the sender's name was the same as the principal's name (in the case of an individual copyright owner), or the sender's organization was the same as the principal organization (in the case of an organizational copyright owner).

[290] These included an individual self-identifying as a "legal agent," several law firms, and a single REO.

[291] Where a sender provided a business name, we categorized the sender as a business. This method is imperfect as individuals that provided a business name are counted as "businesses" using this method, even if it is a self-named sole proprietorship. We used the LexisNexis Academic database to categorize business senders by size. *See Get Company Info*, LexisNexis Academic, https://www.lexisnexis.com/hottopics/lnacademic (last visited Feb. 5, 2016).

medium" businesses of 100 to 999 employees (.7% of the non-Miller requests) or "medium" businesses with 1000 to 9999 employees (.2% of the non-Miller requests). In nearly a quarter (24%) of these requests, the business size could not be categorized—most likely, these are also very small businesses, but we cannot be sure.[292] Excluding Miller's requests, businesses sent nearly two-thirds (66.1%) of requests and individuals sent just over one-third (33.9%) of the requests. Including Miller's requests, individual requests rise to 68.9%. *See Fig.* 9, below.

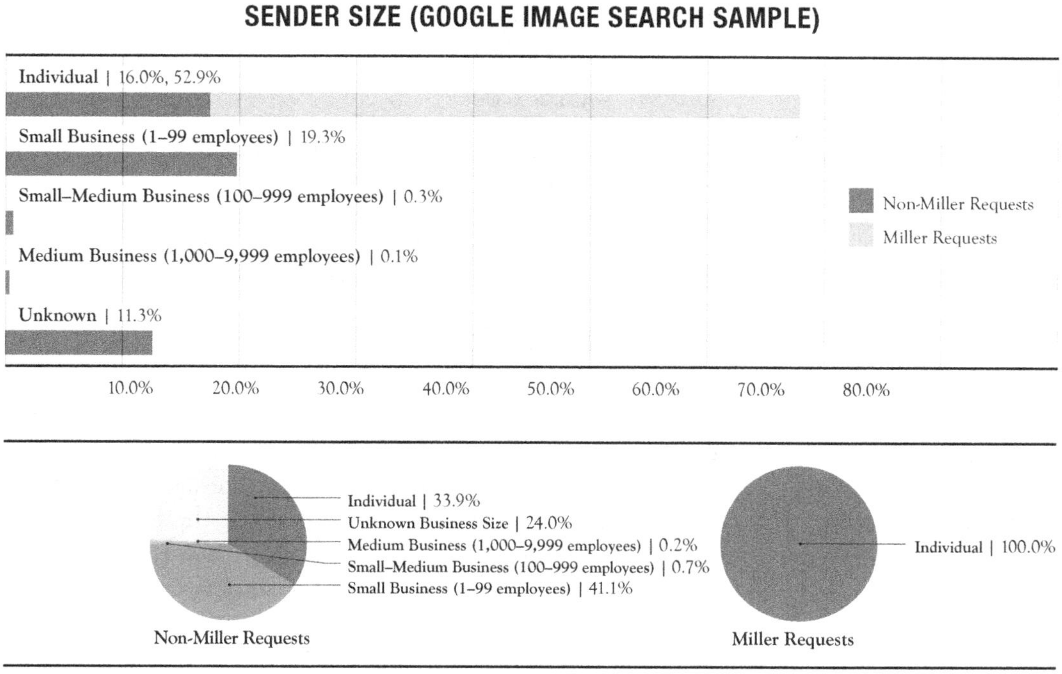

Figure 9: Sender Size (Google Image Search Sample)

Overall, individuals and small businesses sent three-quarters (75%) of the non-Miller Google Image Search requests (increasing to 88.2% when the Miller notices are included). *See Fig. 9*, above. It appears that large, well-resourced senders do not focus on Google Image Search—in our dataset at least, takedown requests come from much smaller players.

Though large corporations appeared to be absent from the Image Search tranche, smaller image-focused businesses were well represented. The most prominent industry actors were the art/design[293] and adult entertainment industries, with each representing over one-fifth (23.8% and 20.3% respectively) of the non-Miller requests. The photography industry made a notable showing, with 10.5% of the non-Miller requests. E-commerce sites represented 5.1% of the non-Miller requests, and the industry was unknown or could not be classified in 6.3% of the non-Miller requests. *See Fig. 10*, below.

[292] These businesses could not be categorized because they did not appear in the LexisNexis Academic database *Id.*

[293] A significant number (81.8%) of the requests originating from the art/design industry were for greeting card designs, with one sender (who was the second-most prolific sender after Ella Miller) sending the majority of these requests.

INDUSTRY OF PRINCIPAL (GOOGLE IMAGE SEARCH SAMPLE; NON-MILLER REQUESTS)

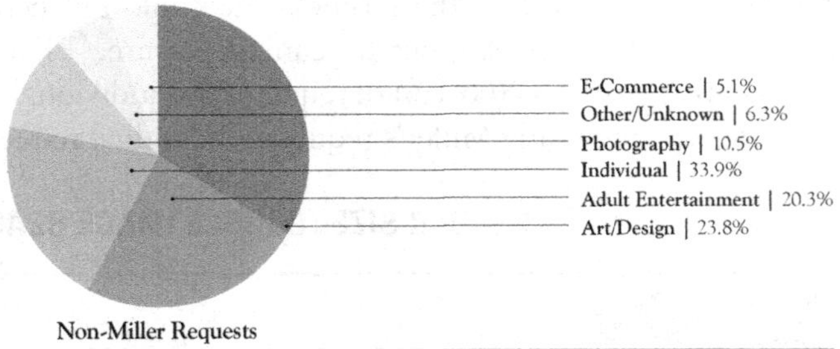

Figure 10: Industry of Principal (Google Image Search Sample; Non-Miller Requests)

As would be expected for an image search service, most senders claim infringement of pictorial and graphic works. *See Appendix* A for details and charts.

iii. Overseas Senders Dominated

In their 2006 study, Urban and Quilter found that 34% of the Google notices they studied targeted material that appeared to reside outside the United States.[294] This raised serious questions about the validity of those notices. The DMCA is a United States law that governs United States entities. Particularly in cases where the parties are outside the United States or the challenged material resides on servers outside the United States requesting takedown based on US copyright law is a questionable enterprise. Yet the dominance of US service providers means that extra-territorial disputes may take place on servers owned by US OSPs. In some cases, these servers may be operated within the United States; in others they may be located in another jurisdiction.

In this study, we did not identify the country where the targeted material resided; however, we did find that overseas senders sent the majority of requests in the Google Image Search notices. All of Ella Miller's requests originate from Sweden, and appear to relate to "flame-wars" between individuals located in Sweden, though some of these arguments may have taken place on servers in the United States.[295] Leaving Miller aside, senders based abroad sent *over half* (56.5%) of the notices to Google Image Search. This included senders based in Germany (16.4%), India (7.5%), Israel (7.5%), Great Britain (3.3%), and China (2.6%). Other foreign countries each had a smaller share but combined made up 19.2%. Senders based in the United States sent 43.5% of the requests (dropping to 20.5% if the Miller notices are included). See *Figs. 11* and *12*, below.

---

[294] Urban & Quilter, *supra* note 5, at 676.

[295] Much of the targeted content appeared on a Flashback discussion forum, which is apparently based in the United States. *Flashback Media Group*, WIKIPEDIA, https://en.wikipedia.org/wiki/Flashback_Media_Group (last updated Nov. 5, 2015).

SENDER'S COUNTRY (GOOGLE IMAGE SEARCH SAMPLE; NON-MILLER REQUESTS)

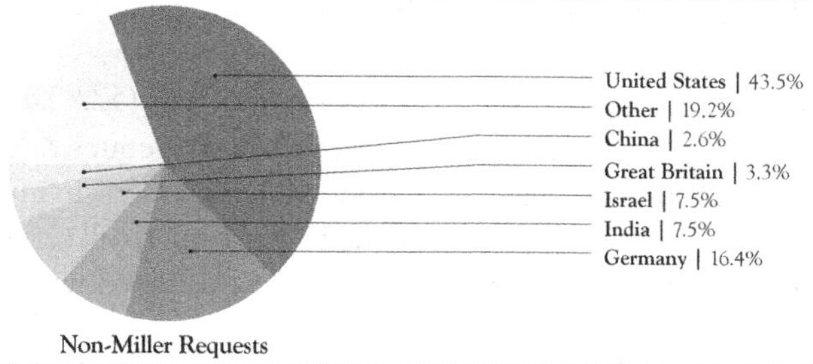

Figure 11: Sender's Country (Google Image Search Sample; non-Miller Requests)

SENDER'S COUNTRY (GOOGLE IMAGE SEARCH SAMPLE; INCLUDING MILLER REQUESTS)

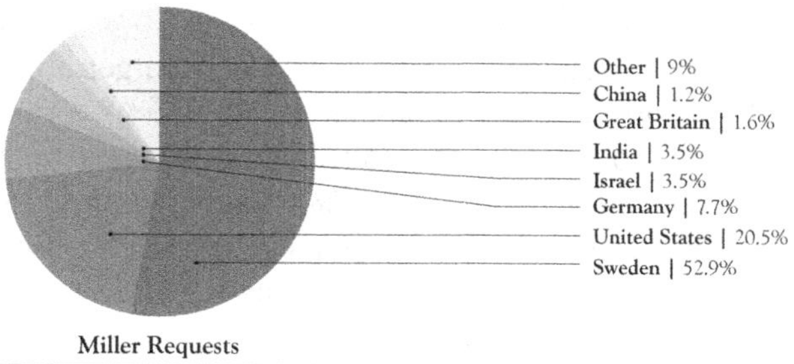

Figure 12: Sender's Country (Google Image Search Sample; including Miller requests)

iv. High-Volume Senders: Ella Miller and Six Others

As with Google Web Search, Google Image Search requests are also dominated by a few prolific senders: Seventy-eight point three percent were sent by only seven senders (out of a total of 197 senders in the coded sample). Their profile, however, is very different from the professionalized Google Web Search senders. Most are individuals or small businesses. See *Table 3*, below.

SENDER	SENDER SIZE	PERCENTAGE OF REQUESTS IN GOOGLE IMAGE SEARCH CODED SAMPLE
Ella Miller	Individual	52.9%
Lil Duck Duck	Small Business	8.5%
Purzel-Video GmbH	Small Business	6.1%
Anatoli Ivanov[295]	Individual	3.5%
Alpha Sky Productions	Unknown	3.1%
Ragalahari	Small Business	2.9%
Oasis Costumes	Unknown	1.2%

Table 3: Percentage of Requests in Google Image Search Coded Sample

[296] This is a pseudonym.

Though they almost certainly have far fewer resources than the dominant Google Web Search senders, these senders also make significant use of DMCA's takedown measures. Each sent hundreds, and in some cases, thousands, of takedown requests during the six-month period.

As introduced above, Miller alone sent over half (52.9%) of the Google Image Search requests, representing thousands of individual takedown requests. Using the machine-coded data for the entire six-month set, we can see that Miller also sent many requests to Google Web Search, and indeed just how prolific a sender she is: her total to all services in the Lumen dataset over the six months it covers was nearly 1 *million* individual takedown requests in over 10,000 notice forms.[297]

b. Target Site Characteristics: Image Search Notices Largely Target Material on Social Media, Personal Websites, and Blogs

The Google Image Search targets also look much different from the torrent and file search sites that make up two-thirds of target sites in the overall six-month dataset. In the Google Image Search corner of the dataset, the "worst of the worst" give way to social media postings, personal websites,[298] and blogs. A quarter (24.6%) of the non-Miller requests targeted links to material on social media sites, and 15.6% target links to material on personal websites or blogs. News sites (7.4%), e-commerce sites (6.9%), forum/fan sites (6.3%),[299] video streaming sites (6.1%), torrent sites (5.9%), aggregator sites (3.8%), cyberlockers (1.7%), corporate sites (1.6%), file search sites (1.0%), and educational sites made up the remainder of the targeted sites in the non-Miller requests.[300] A large number (47.7%) of targeted sites could not be categorized, typically because the targeted site was no longer live. *See Fig. 13*, below.

When coding, reviewers also noted where it was apparent that an individual user's content was targeted, typically because the request was a direct link to a single post on a message board thread or where the targeted site was obviously a blog that belonged to an individual user. More than one in ten (10.9%) of the non-Miller requests clearly targeted an individual user.

[297] Most of these requests—975,674—were to Web Search, and marked as "legalother2" rather than "DMCA" notices. The remainder were to Google Image Search and marked as "DMCA."

[298] Targeted sites tagged as "personal websites or blogs" may include small businesses, particularly where the business model is one based on in-site advertising revenue. Sites selling goods were instead tagged as "e-commerce" sites.

[299] In an oversight, we did not create separate variable for forums and fan sites. These types of target sites were captured together in one variable, which was used to indicate fan sites *or* forums (used by fans or not). See Appendix C.

[300] Nearly half (47.7%) of the target sites could not be categorized, in most cases because the target site was no longer live.

TYPES OF TARGET SITES (GOOGLE IMAGE SEARCH SAMPLE; NON-MILLER REQUESTS)

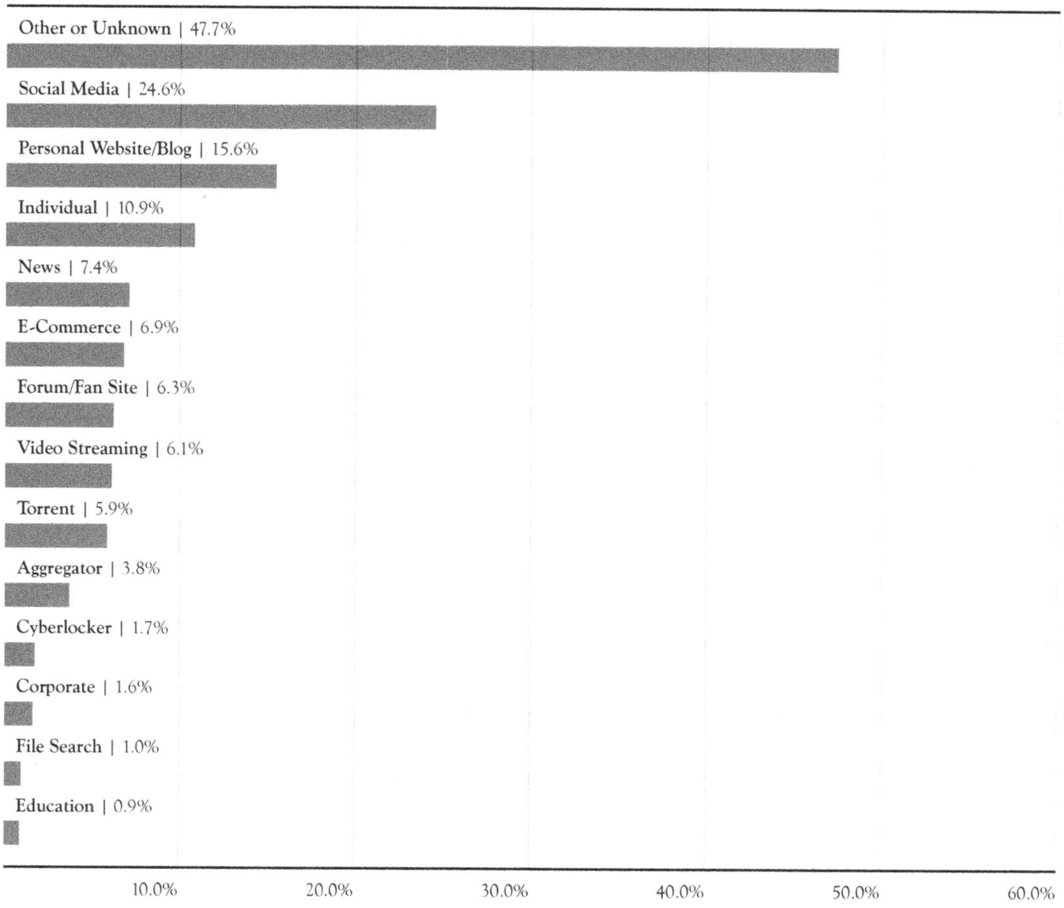

Figure 13: Types of Targets (Google Image Search Sample; Non-Miller Requests)

The prevalence of social media, personal websites, and blogs, and individual user content among the targets raises concerns about the potential impacts on individual expression. When combined with other substantive considerations—such as potential fair use defenses or improper subject matter discussed below—the importance of human review of the claims increases.

c. Miller's Notices Differ from Others but Still Target Individual Content

Miller's notices, which also target the types of sites used by individual users and which raise claims about offensive or contested speech rather than copyright infringement, also raise concerns about their impact on individual expression. In other ways Miller's notices deviate from the others. The vast majority of Miller's requests (91%) targeted written content. (In 4% of Miller's requests, the type of targeted material could not be determined.) The targets of the Miller requests are most often fan sites or forums rather than social media sites, but they are still sites that attract individual users. Nearly 8 out of 10 of the Miller requests target forums/fan sites (78.6%) and 15.3% target material on a personal website or blog. The remainder of the Miller requests target cyberlockers (2.4%), news sites (1.7%), social media sites (.9%), and a handful had unknown targets (.8%). *See Fig. 14*, below. More than 1 in 7 (14.5%) of the Miller requests clearly targeted an individual user. These features all reflect her attempts to use the notice and takedown process to remove allegedly defamatory and abusive message-board postings.

TYPES OF TARGET SITES (GOOGLE IMAGE SEARCH SAMPLE; MILLER REQUESTS)

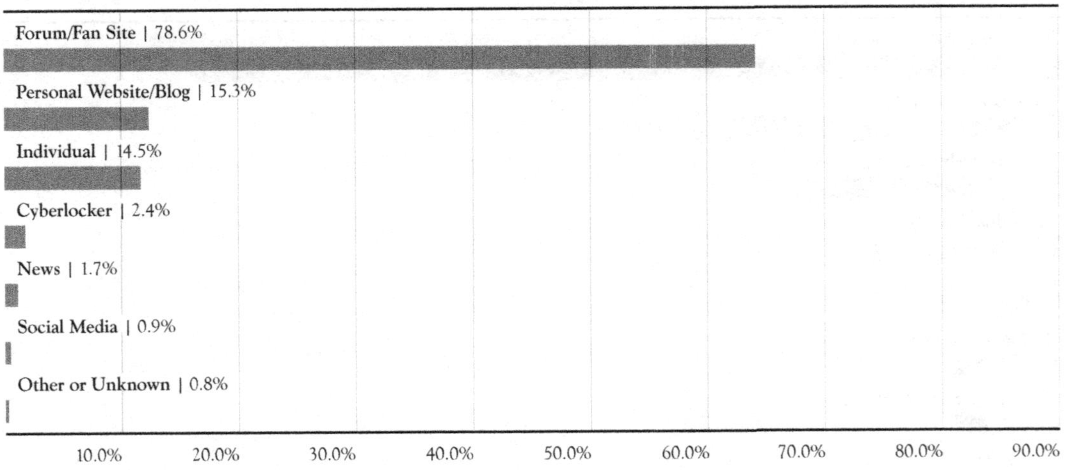

Figure 14: Types of Targets (Google Image Search Sample; Miller Requests)

2. Questions of Accuracy and Substantive Judgment

As in Study 2, we also independently evaluated the Google Image Search requests' substantive compliance with the statute, the strength of the underlying copyright claim, and the appropriateness of the DMCA for addressing the complaint.

The Google Image Search requests presented significantly more potential substantive problems, and a greater variety of potential problems, than the Study 2 notices.[301]

- All of the Google Image Search requests sent by Ella Miller were improper subject matter for DMCA takedown—none were copyright complaints.
- Including Miller's requests, seven out of ten (70%)[302] of the Google Image Search takedown requests presented serious questions about their validity.
- Even without Miller's requests, 36.3% of the remaining Google Image Search takedown requests were questionable. These broke down into several categories:
  - 15.1% raised questions about the subject matter of the claim (this increases to 60% when the Miller notices are included);
  - 11.5% exhibited characteristics that suggested possible fair use defenses;
  - 6.1% presented questions about ownership of the underlying copyright;
  - 2.9% presented questions about whether the sender had properly identified the allegedly infringing material;
  - and a small number (.9%) targeted material likely to be in the public domain.

[301] The substantive problems presented by the Study 3 requests may indeed by present in the Study 2 notices sent by individual senders. However, the requests in Study 2 sent by small senders were swamped by the automated requests. See *supra* note 236.

[302] For purposes of calculating the total number of questionable requests, a request that has multiple questionable characteristics is not counted more than once.

a. All of Miller's Requests Were Likely Invalid

Ella Miller appears to provide an example that supports Study 1 OSPs' "rule of thumb" to more closely scrutinize notices from individual senders than those from professionalized senders. By and large, her Image Search notices do not appear to be proper subject matter for DMCA takedown. Though she identified them as DMCA notices[303]—indicating a copyright claim—by and large they complain of other issues. Usually, they identify links to written content that she alleges is defamatory, harassing, slanderous, or threatening as allegedly infringing material. Though we could not discern details, the notices typically targeted message board threads and blog posts that appear to be critical of Miller. An additional handful of requests identify the AIM as a photograph of Miller. However, these requests are still grounded in defamation and similar torts rather than copyright.

All of Ella Miller's requests thus fall into the "questionable" notice category. While they may be sympathetic complaints, they do not present copyright issues, and thus are not properly addressed through the DMCA. The few requests that identify a photograph as the allegedly infringing material request takedown of photographs that are professional portraits of Miller herself. This raises questions about whether she in fact owned the copyright to the images—which would more commonly accrue to the photographer—or was authorized to send notices on behalf of the copyright owner.

b. The Rest of the Image Search Requests Also Presented Significant Substantive Questions

Leaving Miller aside, 36.3% of the Google Image Search requests presented substantive problems. These broke down into a few categories: questionable subject matter for a takedown request; a potential fair use defense; questions surrounding ownership of the allegedly infringed work; and a handful of less-frequent issues. See *Figure 15*, below.

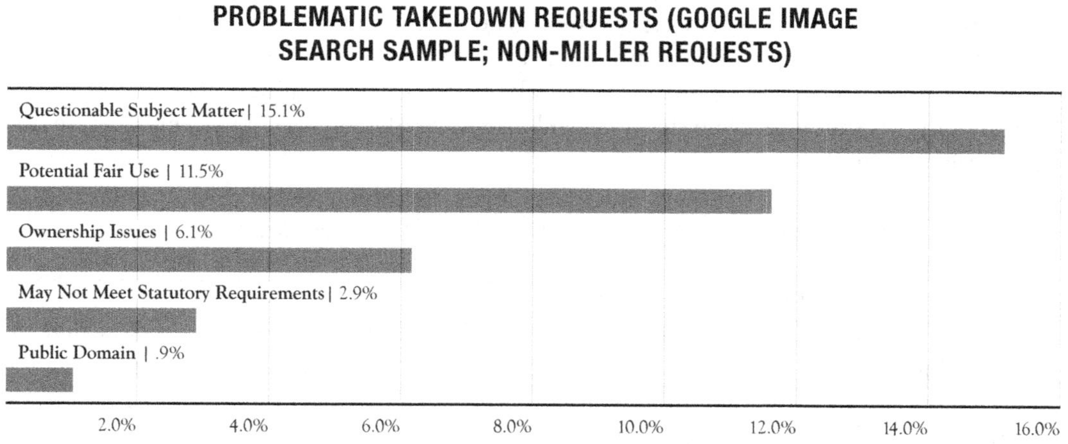

Figure 15: Problematic Takedown Requests (Google Image Search Sample; Non-Miller Requests)

[303] Senders who use Google's web form for Image Search answer a series of questions that are intended to filter requests into the appropriate complaint track. The notices are then labeled with an "issue type" that identifies the type of complaint. See *Removing Content From Google*, https://support.google.com/legal/troubleshooter/1114905?hl=en#ts=1115648 (last visited Feb. 5, 2016).

i. Nearly One in Six Requests Raised Questions About the Subject Matter of the Claim

Like Miller's requests, a significant number—15.1%, close to one in six—of other senders' requests also raised issues outside of copyright, raising questions about the use of the DMCA takedown process. The specific issues raised vary. Most prominent were privacy concerns—typically, when senders appeared in a photograph that they wanted removed from the Google Image Search index. Over one-third of notices with subject matter problems raised privacy concerns (for a total of 6.1% of the non-Miller requests).

Another third of subject matter issues involved thinly protected product photographs. Typically, the sender appeared to be a competitor of the target... Cases like these present an OSP with a potentially difficult judgment call.

As Urban and Quilter saw in their 2006 study, another third of these requests (6%, about one in seventeen of all the non-Miller notices) involved product photographs. Any copyright interest in a product photo is likely to be thin: because they may lack the requisite originality to merit copyright protection, product photos raise questions about the copyrightability of the original work. As most photos involve at least some minimal creativity,[304] the more direct issue may be fair use. This is especially salient as requests to remove product photos typically involved a sender and a target competing in the same line of business. Cases like these present an OSP with a difficult judgment call.

A number of other issues rounded out the subject-matter category. A small percentage of requests raised trademark concerns (2.1%), including a few where the work alleged to be infringed was a short trademarked phrase, such as "Point Master" or "Mosquito Genie." Others identified the allegedly infringed work as visual elements of the trademark or logo but expressed concerns unrelated to copyright. For example, in one request, a French theme park company wanted the identified material removed because it had recently updated its branding and the AIM was "not consistent with our current brand."[305] Finally, a small percentage of requests raised defamation claims (1.5%) and concerns with "online impersonation" and "harassment" (1.3%), indicating that Ella Miller's notices demonstrate problems that are not necessarily typical.

ii. One in Nine Requests Presented a Fair Use Question

Over half of takedown requests that raised potential fair use issues were directed toward material on news sites.

Significantly more potential fair use issues turned up in the Google Image Search notices than in the overall six-month dataset. One in nine of the non-Miller Google Image Search requests (11.5%) were flagged with characteristics that weigh favorably toward fair use, suggesting that further review could reveal a fair use defense. Over half of these were requests to take down allegedly infringing material on news sites. Others included requests where the allegedly infringing material was apparently being used for educational purposes, such as a scientific photograph of bacteria under a

[304] For example, courts have held that sufficient creativity for at least thin copyright protection may come from creative use of lighting, angles, or composition.

[305] Translated to English from original French text.

microscope or photographs of architectural and historical sites in Israel. In some cases, the allegedly infringing material was a part of a curated collection of images. Examples included greeting card designs that were used as examples of "cute invitations" or "wording for a 60th anniversary card." Other requests in this category exhibited a variety of other characteristics that suggested considering fair use would be warranted; for example, cases where only a small amount of the original work was copied, or where the allegedly infringing material adapted, commented on, or criticized the original work.

As with the Study 2 sample, we could not conduct a full fair use analysis. As such, the strength of potential fair use claims will vary.

iii. Ownership Issues

In 6.1% of the non-Miller Google Image Search Requests, the allegedly infringed work was a photograph in which the sender was the subject of the photograph.

In nearly one out of sixteen of the non-Miller Google Image Search Requests (6.1%), the allegedly infringed work was a photograph in which the sender was the subject of the photograph. Since copyright ownership typically vests with the photographer, the subject of the photograph does not usually own the copyright to the photograph in which they appear. Accordingly, in such cases, it is unclear whether the sender has the authority to send a takedown request. There is a recent caveat: in the age of the "selfie" stick and front-facing camera phones, it is possible that the subject of the photograph may also be the photographer and, as such, the copyright owner.[306] Excluding instances where the complained-of image appeared to be a self-portrait drops the number of notices with ownership issues down to about one in eighteen (5.6%).

iv. Only a Small Percentage of Image Search Requests Failed to Identify Clearly the Works in Question

Interestingly, Google Image Search notices presented far fewer problems with identifying the works in question than the Study 2 notices (2.9%). A relatively small percentage of requests linked to search aggregator pages,[307] leading to problems with identifying the allegedly infringing material on the linked page, similar to issues detailed in Study 2. A negligible number failed to identify the allegedly infringed work.

This difference is worth noting; it may relate to the much smaller percentage of Study 3 notices that appeared to be automatically generated relative to the Study 2 notices. It may also relate to the type of content at issue in these requests; visual content is not as likely to be found in the search-result or content-aggregator page types characteristic of the audio-visual and audio content in Study 2 and as such may be less likely to present problems with the identification of the allegedly infringing material.

[306] It is also possible that, in some cases, the photographer transferred copyright ownership to the subject of the photograph.

[307] Aggregator sites curate and present multiple options for accessing content.

v. Public Domain

A small number (.9%) of requests targeted material potentially in the public domain. This set was within the margin of error, so it is merely anecdotal. However, it included some interesting examples, including several drawings of Turkish costumes first published in the early 1800s, and a government-authored work (a mugshot) that also raised privacy concerns.

3. Study 3: Discussion

Discussions about notice and takedown rightly include debates over who can meaningfully use the system. Notices are much cheaper than lawsuits, but that does not necessarily mean that that notice and takedown works well for less-well-resourced copyright owners. Indeed, the notice and takedown process has been publicly criticized by independent photographers, authors, and musicians for requiring too much of smaller copyright owners with relatively limited enforcement resources.[308] The Google Image Search tranche allows us to explore this issue because it reveals an aspect of the DMCA takedown system not readily apparent in the Study 2 sample: extensive use of the system by individuals and other small, non-professionalized senders.

If there are barriers to individuals and small businesses using notice and takedown, they are not apparent in the Google Image Search data. Individuals and small businesses made substantial use of the takedown system. While it is surely more difficult for individuals to exercise notice and takedown on the scale of the automated systems used by REOs, the Miller data suggests that this is not out of reach. The Miller requests provide a potent example of concerns voiced by some DMCA Classic OSPs in Study 1 that even one determined sender could send enough requests to compromise their ability to conduct substantive review.

In general, the types of senders, works, and targeted sites in Study 3 are much more heterogeneous than in Study 2. Because of the idiosyncratic nature of many of these requests, OSP responses may require more difficult, time-consuming, and individualized consideration. Given the likelihood of mistakes, these types of requests seem to be poor candidates for automation; efforts to improve efficiency may be better focused on front-end educational efforts directed at senders rather than automated processing.

Perhaps related to the less-sophisticated nature of typical senders, mistakes and misuse were a disappointingly strong feature of the Study 3 notices. Miller's notices are a prominent example, but she was by no means the only sender incorrectly using the DMCA notice and takedown mechanisms to address non-copyright complaints.[309] Other requests also

---

[308] *See, e.g.*, *Hearing on Section 512 of Title 17*, *supra* note 86, at 54 (2014) (statement of Maria Schneider, Grammy Award Winning Composer/Conductor/Producer, Member of the Board of Governors, New York Chapter of the Recording Academy) (stating that she "must spend countless hours trying to take [her illegally uploaded music] down, mostly unsuccessfully."); The American Association of Independent Music (A2IM), Response to Notice of Inquiry on "Department of Commerce Green Paper, Copyright Policy, Creativity, and Innovation in the Internet Economy," at 3 (Nov. 13, 2013), http://www.ntia.doc.gov/files/ntia/american_association_of_independent_music_comments.pdf (stating that small and medium-sized music enterprises do not have the financial means or resources to engage in widespread copyright monitoring on the Internet); Directors Guild of America, Response to Notice of Inquiry on "Department of Commerce Green Paper, Copyright Policy, Creativity, and Innovation in the Internet Economy," at 3 (Jan. 17, 2014), http://www.ntia.doc.gov/files/ntia/directors_guild_of_america_post-meeting_comments.pdf (stating that notice and takedown places an enormous burden on independent directors who are also copyright holders, and that these individuals lack the resources to monitor and provide notice to websites).

[309] For a high-profile example, see Garcia v. Google, Inc., No. CV 12-57302, (9th Cir. May 18, 2015).

sought to address non-copyright concerns such as privacy and defamation—concerns more appropriately channeled through non-DMCA tracks. Google Image Search requests also showed significant problems with the statutory requirements, requesting takedown where there were potential fair use defenses, ownership issues, and others. A high proportion—more than half—also related to complaints originating outside the United States, bringing into question whether these are appropriate requests, and raising broader questions about the appropriateness of US copyright law for dealing with non-US disputes.

For OSPs that attract takedown notices from individuals and other small, non-professionalized senders, the need for human review of requests is amplified. Human review of requests is crucial to preserving the integrity of open, online platforms that host individual expression, particularly in the face of such frequent instances of mistaken notices. However, the capacity of OSPs to undertake human review can easily be compromised by growth in the use of the system—and if this study is representative, even by a single, determined sender.

Where Study 2 identified some potential problems with automation, this study reveals a very different set of challenges. Smaller senders need better information about when takedown notices are appropriate, and better sending practices. OSPs need good ways to review and manage notices from these types of senders. Targets need meaningful ways to address problematic takedown requests. Our recommendations in Section V thus suggest best practices and reforms that address some of the issues affecting individuals and other small, non-professionalized senders as well as those arising from automated systems.

V. ANALYSIS AND RECOMMENDATIONS

Debates over online copyright infringement and how to address it are vociferous. For understandable reasons, they tend to focus on valuable copyrights, large-scale infringement, and major intermediaries—such as Google, or connectivity provider ISPs—that touch broad swaths of the Internet. Yet our research shows clearly that this is not a sufficient account of the challenges of balancing copyright, freedom of expression, and capacities to innovate online. DMCA Classic notice and takedown is still the most common practice at OSPs, and, for most, it is still sufficient to manage the takedown requests they receive. Some of these DMCA Classic OSPs fear the arrival of floods of automated notices or requirements to implement expensive DMCA Plus measures that seem unnecessary and that they cannot afford. In their view, such developments would undermine crucial safe harbor protection, damage freedom of expression, and, in some cases, increase competitive advantages for more well-resourced market participants. Policy interventions should take this set of interests into account.

Senders are similarly split. While we were unable to directly explore the experiences of smaller senders in this set of studies, both the Google Image Search notices we reviewed in Study 3 and OSPs' experience with less professionalized senders strongly suggest that policy decisions based on the knowledge and actions of large rightsholders and REOs will leave out important challenges and issues specific to small senders and their targets.[310] Unfortunately, Study 3 also suggests that less knowledgeable senders may send a significant number of problematic takedown requests.

Finally, interviews and surveys made clear that OSPs often operate without much insight into others' practices. These information asymmetries exacerbate worries and misapprehensions about the reasons behind other market players' situations.

Our research maps this complex ecosystem, with its distinct challenges for large incumbents and smaller players, for those in the center of the "copyright wars" and those on the edges, and for those playing different roles in the technological ecosystem. The process described in section 512 is simple and one-size-fits-most, while on-the-ground practice is bespoke, and anything but simple. While the underpinning structure—relatively easy takedown for copyright holders, and a safe harbor for OSPs—remains basically intact, changing technology, evolving business models, and the growth of Web 2.0 incumbents are stressing the system for some OSPs and rightsholders. In this section, we analyze the current state of notice and takedown and proposals to extend or change it, and offer recommendations for legal and practice reforms that take into the complexity of notice and takedown as currently practiced.

[310] Although a limited number of individual rightsholders and smaller OSPs have participated in multi-stakeholder forums and congressional hearings, our research suggests that a far broader group interacts with the notice and takedown system than is represented in these public forums. For example, the individual rightsholders represented in these forums does not begin to cover the diversity suggested by Study 3. Further, some OSPs reported that they did not have the resources to follow or participate in these debates.

A. NOTICE AND TAKEDOWN'S SUCCESSES

Notice and takedown has not enjoyed the best press in recent years. Amid a steady stream of anecdotal reports of abusive takedowns,[311] complaints by large rightsholders that determined online pirates continue to dent their profits,[312] and updates to Google's Transparency Report—showing what seems to be an ever-escalating arms race fought with millions of automated notices and revolving offshore domains—it would be easy to conclude that notice and takedown is practically obsolete.[313] The enduring importance of notice and takedown to both OSPs and rightsholders may thus seem surprising. And indeed, the system exhibits important failures and definite strain, which we discuss below. But in some of its most basic features, the notice and takedown system is functioning, and meets the goals it was intended to address.

First, it is hard to overstate the importance of the section 512 safe harbor to OSPs. Because United States copyright law provides for injunctions and very high statutory damages, calculated per infringed work,[314] platform providers view copyright risk as impossibly high without the safe harbor. All stressed that the safe harbor is both central to their ability to provide their services and "baked in" to the structure of the online ecosystem. The proliferation of services both personal and enterprise-level—website hosting, email services, blogs, social networking, fan sites, photo and video platforms, distributed "cloud" storage and computation, and many others—appears to be possible because of the safe harbor from secondary copyright liability that section 512 provides.

Moreover, if it is to protect OSP market entry and competition, the safe harbor has to be more than a legal fiction; it must be practically available. OSPs are acutely aware of this. In Study 1 OSPs reported fear of its practical disappearance if soft pressures to adopt expensive DMCA Plus measures such as filtering are hardened, pushing DMCA Classic OSPs away from traditional notice and takedown and into direct competition with much more well-resourced incumbents that employ heavily automated systems. We discuss this further below.

Second, notice and takedown continues to provide an efficient method of enforcement in many circumstances—especially compared to lawsuits. In Study 1, major rightsholders generally agreed that notice and takedown is a main tool for addressing online infringement, and most described it as having some success in managing (though certainly not eradicating) infringement on DMCA-compliant sites. Studies 2 and 3 also showed extensive use of takedown by a wide variety of rightsholders, including smaller copyright owners, though the picture is necessarily limited by the fact that most notices targeted links controlled by one large provider (Google). Overall, major rightsholders described strategic deployment of automated noticing systems, together with user education, as bringing some meaningful success in enforcing copyright on DMCA-compliant sites. (Extra-territorial infringement-focused sites, however, are another matter; they are discussed under "limitations" below.)

---

311 For some examples of abusive takedown requests, see *Takedown Hall of Shame*, ELEC. FRONTIER FOUND., https://www.eff.org/takedowns (last visited Feb. 5, 2016).

312 *See, e.g., Who Music Theft Hurts*, RIAA, http://www.riaa.com/physicalpiracy.php?content_selector=piracy_details_online (last visited Nov. 10, 2015).

313 Academic literature, too, has regularly catalogued notice and takedown's challenges and failures. *See supra* note 5.

314 17 U.S.C. § 504(c) (2012); Pamela Samuelson and Tara Wheatland, *Statutory Damages in Copyright Law: A Remedy in Need of Reform*, 51 WM. & MARY L. REV. 439 (2009), http://scholarship.law.wm.edu/wmlr/vol51/iss2/5.

We discuss some of the strains large-scale infringement and large-scale automated noticing create in the next section.

In certain cases, notice and takedown appears to work on a procedural level, too. The process remains an effective mediator of conflicts between rightsholders, OSPs, and users when conducted on a small scale, bookended by human review on the part of both knowledgeable, good-faith senders and OSPs. Moreover, because the counter notice process remains underused by targets, due process success depends heavily on OSPs' ability to undertake substantive review of notices' merits, and importantly, their willingness to reject problematic notices. While we were only able to independently review OSP practices for notices available through Lumen, most respondents gave descriptions of a process that involves some substantive review. And though public accounts are still few, the rejection rates published by a range of DMCA Classic OSPs—some of which exceeded 50%[315]—also suggest that it is possible to conduct substantive review if the scale of notices received remains manageable. For OSPs outside the main copyright conflict zones around search, music, and video services, such practices appear to remain the norm.

There are several important caveats. First, transparency into OSP practices is still very limited, making it impossible to judge how well notice and takedown works procedurally across the online ecosystem. Second, successful notice and takedown relies on knowledgeable senders acting in good faith. However, the high numbers of questionable notices we saw in Study 3 suggest that senders outside the professionalized creative industries may make multiple mistakes, increasing liability and resource pressures on OSPs that, combined with the weakness of the counter notice process, may be too great to avoid overbroad takedown. Third, and relatedly, our Study 2 and 3 findings show that significant numbers of problematic notices are likely to get through if the scale of noticing rises beyond some threshold. Automation poses a significant risk of over-inclusiveness, especially if not backed up by human review. Finally, OSPs' oft-expressed fears of losing the safe harbor lead them to make conservative decisions, biasing them toward takedown. They told us this in Study 1, and our Study 3 numbers—where 70.3% of notices exhibited validity questions, though Google removed 58.8% of the complained-of links[316]—tend to support this. (These numbers, however, also suggest that large-scale decision-making, if well implemented, can weed out a meaningful portion of problematic notices.) Accordingly, we might expect that even the 50-percent-plus rejection numbers reported by other DMCA Classic OSPs may reflect conservative decisions.

Overall, the fundamental compromise in section 512—to manage liability and enforcement costs for OSPs and rightsholders—holds in essence. The basic compromise still underpins negotiations between OSPs and rightsholders over responsibility as Internet services and distribution channels evolve. Still, notice and takedown's limitations—and in some important regards, apparent failures—are also significant, and begin to threaten the fundamental compromise. We now turn to these.

---

[315] *See supra* Section III.B.2.

[316] E-mail from Michael Deamer, Legal Assistant, Google Inc. (Aug. 22, 2014, 12:13 PST) (on file with authors).

B. NOTICE AND TAKEDOWN'S LIMITATIONS AND FAILURES

In practice, notice and takedown presents significant problems. Because of their negative effects on online expression and competition, substantive mistakes that lead to questionable takedown are perhaps most concerning. In our studies, these primarily cropped up in two scenarios. First, Study 2 showed that the automated tools used by some rightsholders are too broad and have insufficient safeguards, resulting in questionable requests in nearly a third of cases. Second, in Study 3, requests sent by less-sophisticated notice senders raised a sobering number of questions about the appropriateness of the underlying requests. Thus, there is a two-fold problem that splits between sophisticated and less-knowledgeable senders. In the first instance, machine-based decision-making appears to require more and better review by knowledgeable humans. In the second, human fallibility appears to be the main issue. We suspect the latter is closely related to limitations in these senders' knowledge of copyright law and sense of when takedown is appropriate.

1. Mistaken or Questionable Removal of Content

a. Problems of Automation: Mistaken Takedowns and Pressures on Due Process

The rise of mass notice sending via automated systems raises immediate questions of accuracy and due process. Human scrutiny of underlying claims necessarily decreases when by-hand infringement detection, noticing, and review are replaced by automated systems. Understanding how this may affect the accuracy of takedowns was a major question in our research.

We found reason to be concerned when human review is replaced with a high degree of automation. The automated notices we examined in Study 2 were, in the main, sent by sophisticated rightsholders (or their agents) with a strong knowledge of copyright law, yet nearly a third of the notices raised questions about their validity, and one in twenty-five apparently targeted the wrong material entirely.

This observation occurred despite the fact that, in Study 1, both rightsholders and DMCA Plus OSPs described using a variety of checks in an attempt to avoid automation mistakes. Rightsholders described profiling targeted sites before sending notices, subjecting notices to human review if "someone writes back to say what they are doing is fine," and in some cases, periodic spot checks. Such efforts surely help. But the persistence of these types of problems in Study 2 suggests that there is need for better checks to make automated algorithms as accurate as possible, and for more human review at the sending stage of the process.

Appropriate use of automated systems also requires weeding out overly aggressive or bad-faith senders. We spoke only with reputable senders in Study 1, and the notices we reviewed in Study 2 were also, by and large, sent by or on behalf of known, reputable rightsholders targeting file-sharing sites that rightsholders consider "dedicated to infringement." Our respondent senders—including REOs—stressed the importance of reasonable measures for avoiding mistakes. Unfortunately, interviews with OSPs made it clear that the absence of effective liability for bad requests means that there is no pressure on senders to adopt such measures. Senders, accordingly, vary widely in their behavior and degree of concern for improper takedown. In the worst cases, DMCA Auto and DMCA Plus systems that facilitate bulk notice sending and streamline removal may facilitate bad-faith, abusive practices.

For example, one OSP respondent who had caught the attention of a well-known "copyright troll," described receiving massive numbers of "trash" notices, apparently designed to make it difficult or impossible to respond. While this OSP used automated measures to manage notices, it could not readily address this type of bad-faith activity. The complement to large-scale fusillades is "hiding in plain sight"—if properly formatted, but bad-faith, requests for takedown lurk within the thousands or millions of notices directed at file-sharing, automated systems may be hard-pressed to detect them. Indeed, the same OSP who had caught the attention of the "copyright troll" automatically removed content in response to a complaint—an understandable approach to large numbers of notices that nonetheless leaves this risk in place.

This situation creates a conundrum. Rightsholders that focus their efforts on the file-sharing sites, combined with general rules to minimize errors, lower the risk of significant effect on freedom of expression. But our research shows that mistakes are still likely. This, combined with incentives for bad-faith actors to exploit automated detection systems, leaves substantial questions about due process for targets and attendant effects on expression. In section V.D. below, we suggest some reforms and best practices to help address this conundrum.

b. Substantive Mistakes and Abuse: Human Fallibility

Human attention in notice sending is not a panacea, especially in the case of non-professionalized senders. Around seven out of ten of the Study 3 requests—nearly all sent by individuals or small businesses, and most apparently sent by hand—presented serious questions about their validity, ranging from complaints not grounded in copyright, to potential fair use defenses, to a range of other substantive problems.

These findings intensify concerns raised in Study 1 about the quality of notices sent by individuals or one-off senders: a common theme among OSP respondents was that these notices received heightened scrutiny because less-experienced senders are the most likely to misunderstand the notice and takedown process or use it for improper purposes. The high number of problems in the less-professionalized Study 3 notices raises serious questions about how well less-sophisticated, average senders understand copyright law (or alternatively, care about its integrity when they want something removed).

As the small-scale counter notice abuse we saw in Study 1 shows, determined abuse, like determined infringement, can never wholly be contained. Given the numbers of issues we saw in Study 3, however, we expect that better understanding of copyright law and takedown could help. And even intentional abuse can be disincentivized and to some degree, remedied. We make suggestions along both lines below in section V.D.

2. Due Process Failures for Targets

The identification mistakes, substantive problems with claims, and other issues we observed confirm the prescience of Senator Ashcroft's worries, during the drafting of the DMCA, about targets' due process and expression rights. However, the attempted solution—the counter notice and putback process—has been repeatedly criticized as insufficient.[317] Both our qualitative and quantitative studies lend support to these concerns, and indeed suggest that section 512's approach to due process for targeted users is one of its major failures.

---

[317] *See, e.g.*, Alfred C. Yen, *Internet Service Provider Liability for Subscriber Copyright Infringement, Enterprise Liability, and the First Amendment*, 88 Geo. L.J. 1833 (2000); Emily Zarins, *Notice Versus Knowledge Under the DMCA's Safe Harbors*, 92 Cal. L. Rev. 257, 291-95 (2004).

As a procedural matter, material that is targeted by a takedown request is often removed before the target is given the opportunity to respond; this was confirmed in interviews with OSPs and rightsholders. Yet all available evidence suggests that counter notices are simply not used. It is indicative of the problem that the most memorable uses of counter notices for our rightsholder respondents were a few bad-faith, bogus counter notices from overseas pirates. Given the high numbers of apparently unchallenged takedown mistakes that showed up in our quantitative studies, we would expect to see higher numbers of appropriate, good-faith counter notices if the process were working as intended.

Unfortunately, under current practice, there seems to be little chance of this changing. Study 1 OSPs described hesitating to encourage targeted users to send counter notices, even when it seemed appropriate, for fear of creating liability risk for targets and themselves. Unbalanced liability standards—fear of suit by copyright holders but not users—creates incentives for OSPs to take down material. Moreover, some of the main targets of large-scale requests—search services—have no service relationship with targets or any duty to inform them that links are being removed, making it highly unlikely that the target would know to send a counter notice. Further, as we discuss in recommendations, section 512 currently leaves unclear whether search engines are protected for putback like hosting entities, exacerbating the challenge. Overall, the counter-notice process's procedural features make it difficult for OSPs to use it as intended.

The counter notice process contains other flaws. In Study 1, OSPs described it as intimidating and confusing for targets. These issues are compounded by statutory process problems that work against targets' ability to reinstate lawful material in a timely manner. Section 512(g) currently requires OSPs to wait ten to fourteen business days before reinstating material—a very long time in the case of timely commentary and enough in some cases to cause economic loss to the target.

The millions of potential problems we found in the Study 2 and Study 3 requests make these concerns concrete. With anywhere from a third to two-thirds of notices in our samples exhibiting potential problems—and yet no evidence of counter notice responses—the acute defects in the counter notice system become evident. Our samples likely suggest a particularly sharp issue, as targets of takedown requests to search indexes do not receive notice of the requests. Respondents agreed, however, that counter notices are exceedingly rare for hosting services as well. Targets who are victims of takedown misuse or have legitimate defenses appear to be ill served by the well-intentioned counter notice scheme. Clearly, more usable checks are needed.

In the end, counter notice and putback give the appearance of due process for targets without the necessary components of definite notice of the claimed transgression, a reasonably exercisable ability to respond (preferably before action is taken), and an unbiased adjudicator.[318] In the recommendations section below, we build on others' efforts to offer suggestions for improving this situation. Moreover, further expansion of the notice and takedown model, or changes to it, should take into account the fact that targets' expression rights are fragile in a system with strong removal incentives for complainants and intermediaries, but with such limited countervailing incentives to preserve or reinstate improperly targeted speech.

[318] *See* Henry J. Friendly, "*Some Kind of Hearing*," 123 U. PA. L. REV. 1267 (1975).

3. Limits on Ability to Address Large-Scale, Off-Shore Infringement

In assessing the effectiveness of notice and takedown, Study 1 rightsholders drew a sharp distinction between, as one put it, "[OSPs] that believe their business models are legitimate" (even if contested) and businesses with "hardcore institutional models built on piracy." As this respondent described, the latter "operate in the shadows" (and often, offshore), ignoring notices and requests to negotiate. While notice and takedown helps manage copyright infringement by users of most OSPs, rightsholders expressed great frustration with these extra-territorial infringement-focused sites, which are out of the reach of U.S. jurisdiction. Rightsholders' targeting of search engines, like their efforts to enlist payment providers and advertising networks, is an attempt to put pressure, however limited, on those sites.

Several factors limit notice and takedown of infringement-focused sites. One is technical change. The DMCA was passed just before peer-to-peer file sharing arrived on the scene. The one-to-one notice to infringement model is strained by peer-to-peer file sharing, dynamic linking, re-seeding, and other methods used by file sharers and file sharing sites. Automated sending is a means of addressing this more resilient piracy ecosystem, but it is clearly a management tool rather than a comprehensive solution to the problem. As a result, some rightsholders have pushed for requiring DMCA Plus measures like filtering, or even Para DMCA measures such as ISP-level site-blocking.[319]

We are sympathetic to rightsholders' concerns. However, our research also highlighted the fact that many of these suggestions would ill fit the large number of OSPs for which DMCA Classic practices remain adequate, while further exacerbating due process issues for targets. And as discussed further below, changing the safe harbor requirements to include these more expensive measures risks undermining the essential success of notice and takedown in supporting a robust, competitive marketplace for online speech platforms.

These are high potential costs, especially as determined pirates are unlikely to respond to beefed-up legal measures, and are likely to route around stronger technical measures—poorly conceived policy changes could be ineffective, yet cause collateral damage to legitimate OSPs. Instead, we suggest some measures for improving automated tools.

4. Lack of Transparency: Public Policy and Incentives Toward Takedown

For better or for worse, Internet speech relies on the private speech platforms provided by OSPs. Practically speaking, their copyright decisions play a major role in how online expression is regulated. Since the DMCA was passed, however, commentators have voiced concerns about this outsourcing of copyright adjudication to a private process, in which decisions are made not by courts, but within companies.[320] We undertook this research in

---

[319] *See, e.g.*, RIAA, Response to Notice of Inquiry on "Department of Commerce Green Paper, Copyright Policy, Creativity, and Innovation in the Internet Economy" (Nov. 15, 2013), http://www.ntia.doc.gov/files/ntia/recording_industry_association_of_america_comments.pdf; Broadcast Music, Inc., Response to Notice of Inquiry on "Department of Commerce Green Paper, Copyright Policy, Creativity, and Innovation in the Internet Economy" (Nov. 15, 2013), http://www.ntia.doc.gov/files/ntia/bmi_comments.pdf; *see supra* Section III.D. for the reasoning Study 1 rightsholders offered for favoring these techniques.

[320] *See, e.g.*, Dena Chen, Musetta Durkee, Jared Friend, & Jennifer Urban, Updating 17 U.S.C. § 512's Notice and Takedown Procedures for Innovators, Creators, and Consumers (2011). Disclaimer: Public Knowledge's white paper was prepared by Samuelson Law, Technology & Public Policy students, representing Public Knowledge under Urban's direction. The opinions in the white paper belong to Public Knowledge and not necessarily any author of this report, though, based on our research, we endorse some of them here as well.

large part because this private adjudication has produced little public information about whether the notice and takedown system "works" according to the various criteria in play—for example, with regard to the availability of pirated materials on specific sites and in general, or with regard to balancing copyright and speech rights.

Though there are signs of change, few OSPs and rightsholders share information publicly about their notice and takedown practices—either with respect to how they approach notice and takedown overall or how they handle judgment-based decisions. This two-sided embargo constrains public policy discussions about rightsholder and OSP responsibilities, effects on targets, and copyright enforcement more generally. This leaves policymakers working in the dark.

Participants in the notice and takedown system also lack information, and this deficit affects how various actors in the notice and takedown ecosystem perceive risks and implement notice and takedown. Most of our OSP respondents lacked awareness of other actors' practices and reasoning, and their uncertainty encourages conservative behavior that favors takedown.

For example, Study 1 OSPs commonly lacked insight into major rightsholders' reasons for targeting certain OSPs with automated notices or pressure to implement filtering or similar measures. As a result, these OSPs had no way to gauge whether or when rightsholders might turn their attention to the OSP's service and begin sending floods of notices or requesting DMCA Plus measures. These OSPs feared that a sudden change of fate could compromise their ability to substantively review notices, or even raise compliance costs past their ability to manage.

The opacity around takedown practices also stymies improvements in practice. An example from our studies illustrates. In Study 3, more than 70% of notices presented questions about their validity. We were able to verify with Google that it rejected more than 40% of these requests, but no further details. The rejection rate is respectable; it suggests some useful review practices, which may be automated or human or both, from which other OSPs could benefit. The gap between our findings and the rejections also suggests room for improvement. Information-sharing and greater transparency could both help other OSPs better detect problematic notices, and perhaps help Google tighten its Image Search review further.

Some secrecy, of course, is defensible. In Study 1 interviews, rightsholders stressed that releasing some details of their enforcement practices would help pirates develop countermeasures. OSPs, for their part, were most concerned that revealing the details of their takedown practices could subject them to negative attention by rightsholders or targets. This concern tended to beg the question however: it arose from lack of knowledge about whether others' practices differed. OSPs all believed they were complying well with the requirements, but worried that acknowledging difference in practices could create pressure to conform with the most aggressive enforcement practices.

But secret algorithmic decision-making, especially when it affects individual rights, has rightly come under criticism.[321] Both rightsholders' and OSPs' concerns are understandable, but can be addressed. Rightsholders need not reveal their current practices in detail so long as notices are available for review within a reasonable timeframe. OSPs should be assured that their safe harbor protection is not vulnerable, so long as they are complying with the

[321] *See, e.g.*, Danielle Keats Citron & Frank Pasquale, *The Scored Society: Due Process for Automated Predictions*, 89 WASH. L. REV 1 (2014); Edward W. Felten, *A Skeptical View of DRM and Fair Use*, 46 COMM. ACM 56 (2003).

statute. Reasonable privacy for targets can be accomplished through redaction. Finally, while transparency reporting may be prohibitively costly for some due to legacy systems, a measure of transparency can be accomplished in other ways. We offer some suggestions below.

5. Cost and the Safe Harbor's Practical Availability

The cost of implementing notice and takedown was a main theme in Study 1. Whether this becomes a threat to goals behind the safe harbor is unsettled, and depends in part on policy decision-making. The current situation is uneasy. OSPs both described the section 512 safe harbor as fundamental to their freedom to operate and expressed great concern about variable and growing costs. Many OSPs considered the practical availability of the safe harbor to be in jeopardy if costly DMCA Plus measures become requirements. These OSPs, especially those of small or medium size, fear the advent of technical requirements (such as filtering or staydown) that they cannot afford to implement, floods of automated notices that they cannot adequately review for abuse or mistake, and the loss of their practical ability to rely on the safe harbor.

Cost sensitivity ranges along a continuum. At one end are a few very large OSPs with the capacity to spend large sums—sometimes tens of millions of dollars or more—on various enforcement systems; at the other end are early-stage start-ups and very small companies. The capacity to absorb the cost of implementing most automated systems appeared to begin well toward the large OSP end of the continuum. Small and medium-sized OSPs simply did not have the resources. As discussed further below, this would create a competitive disadvantage for most OSPs if compliance requirements were to shift. Good policy decisions would avoid this result; we suggest some guidelines in the recommendations, below.

The cost of identifying infringement and sending notices, especially for small senders, is also a commonly voiced concern. Study 3 suggests that smaller copyright owners can make extensive use of the system Google provides, though managing takedown across a wider array of sites and services likely creates more difficulties for senders with limited resources. However, more research is needed to understand how smaller senders manage infringement discovery and noticing to other OSPs. Strikingly, Study 3 also suggests that smaller senders have significantly more problems with accuracy and inappropriate targeting than larger copyright holders. We make suggestions to help small senders below.

C. NOTICE AND TAKEDOWN'S POTENTIAL FUTURES

1. Proposed Refinements and "Para DMCA" Reformations

Even as notice and takedown has spread beyond US shores and beyond copyright law, scholars, industry stakeholders, and others have offered a steady supply of proposals to address its perceived limitations and failures.[322] These join an ongoing stream of court decisions interpreting the statute. As Internet distribution has grown, rightsholders, especially, have pressed for a move away from noticing and removal of specific infringements and towards both technical and policy features that they view as more efficient for managing large-scale infringement. In our research, OSPs and rightsholders often differed sharply in their assessments of these proposals, and how they might affect the online ecosystem.

---

[322] *See, e.g.*, Chen, Durkee, Friend, & Urban, *supra* note 320; Quilter & Heins, *supra* note 5; Seng, *supra* note 5; Urban & Quilter, *Efficient Process*, *supra* note 5; *Comments Received on Department of Commerce Green Paper*, Nat'l Telecomm. & Info. Admin. (Nov. 15, 2013), http://www.ntia.doc.gov/federal-register-notice/2013/comments-received-department-commerce-green-paper-11132013.

a. Government-Facilitated Best Practices and Voluntary Agreements

The U.S. Department of Commerce's Internet Policy Task Force ("IPTF") recently facilitated multi-stakeholder discussions aimed at identifying best practices or producing voluntary agreements for improving the operation of notice and takedown, without the need for legislative changes.[323] The group ultimately settled on identifying ways to improve the efficiency of the handling and processing of notices by both senders and recipients.

Early discussions explored the possibility of cross-service standardization of the notice intake process. Rightsholder groups, who view the array of idiosyncratic and changing submission systems as an impediment to more efficient notice and takedown, were the primary proponents.[324] Standardization at this level could indeed make notice sending easier for some rightsholders who do not want to customize their notices to adhere to submission formats that vary by OSP. OSPs, however, have been wary of the costs of retooling IT systems, and especially, of formalizing DMCA Auto and DMCA Plus measures like bulk submission processes and "fast lane" processes, especially when the volume of notices the OSP receives does not warrant it. Because such measures are already common among the high-notice-volume OSPs, the primary impact of such standardization would likely be on DMCA Classic OSPs.

OSPs and user groups also expressed concerns about further increasing submissions without corresponding efforts to improve their quality. OSP and user representatives' proposals accordingly focused on quality measures, such as stronger attestations of good faith (to deter abuse of the system), required statements that remind submitters to consider fair use or other limitations, and clearer specifications regarding the identification of content. Rightsholders generally characterized such measures as additional burdens. The RIAA, for example, argued that statements about the liability associated with the misrepresentation of copyright claims (which both Google and Microsoft, for example, include in their submission forms) potentially discourage claimants.[325]

The multi-stakeholder work reflects the complexity inherent in trying to standardize approaches, given the very different profiles of OSP and rightsholder participants. It has proven difficult for the stakeholders involved to come to agreement on what might initially seem to be fairly innocuous efficiencies. In the end, OSPs, user representatives, and rightsholders agreed on a more flexible list of "good, bad, and situational practices" for the notice and takedown process.[326]

---

[323] *Copyright Policy*, U.S. Patent & Trademark Office, http://www.uspto.gov/learning-and-resources/ip-policy/copyright-policy (last updated July 28, 2015) (see subsection on "Multistakeholder Forum on the DMCA Notice and Takedown System" under "2015 Meetings and Comments").

[324] RIAA, Presentation for the Department of Commerce's Internet Policy Task Force Second Public Meeting (2014), http://www.uspto.gov/ip/global/copyrights/RIAAv2.ppt (stakeholder submission from the RIAA to the Department of Commerce).

[325] *Id.* Notices to Google Search from individual senders rose more than seven-fold between 2009 and 2012, suggesting that this concern may not hold.

[326] Department of Commerce DMCA Multistakeholder Forum: DMCA Notice-and-Takedown Processes: List of Good, Bad, and Situational Practices, http://www.uspto.gov/sites/default/files/documents/DMCA_Good_Bad_and_Situational_Practices_Document-FINAL.pdf.

b. Filtering, Hash-matching, and Staydown

DMCA Plus measures that move beyond removal of specific, identified infringements to *ex ante* filtering or "staydown" were a flashpoint in our interviews. Rightsholders have regularly, though unsuccessfully, pressed the issue in courts, arguing that gaining the statutory safe harbor requires more than responding to separately identified infringements. Both filtering and its expansion to "staydown"—in which the same work is thereafter kept off the OSP's platform—were spoken of approvingly by rightsholders. But these measures were deeply controversial in the OSP community. And all OSPs feared potential changes in legal rules or norms that would turn such measures from voluntary into requirements. Indeed, this is one of the most consistent themes we heard.

In practice, the adoption of filtering by a number of user-generated-content-based services, such as YouTube, SoundCloud, and Vimeo, represents a profound shift in the practice of enforcement from takedown based on rightsholder identification of unauthorized uses. For OSPs managing large-scale automated notices, filtering and related measures were generally put in place at considerable expense and after considerable pressure from rightsholders. Though many of the OSPs we interviewed had not experienced direct pressure to shift from DMCA Classic measures (or DMCA Auto measures centered simply on managing large numbers of notices) to DMCA Plus filtering, site-wide takedown via hash-matching, or "staydown," managing this pressure was a major issue for those that had.

In interviews, rightsholders often based their assessments on whether an OSP is perceived as operating a "legitimate" service or primarily trading in pirated content. Although the debate is framed as if such labels are obvious, they can be hard to apply in the zone between offshore pirate sites and more traditional platforms. At times, rightsholders' distinctions appear to turn on an OSP's monetary investment in DMCA Plus measures and its general willingness to relax the requirement to target only specific infringements. In a typical example, one rightsholder listed the following items as indicators that a storage OSP was likely to be legitimate: termination of repeat infringers, fingerprinting, limits on upload bandwidth, and restrictions on public links to the stored material. Only the first is a requirement for safe harbor protection under the DMCA.

Others spoke approvingly of the monetization features that can be built into fingerprinting and filtering systems. Such features suggest a possible solution for rightsholders reeling from the disruption of classical distribution models. But they also raise well-documented concerns about fair use and prior restraint of speech, as well as concerns from follow-on creators about control and fair compensation.

Caught in the middle are compliant OSPs that cannot afford DMCA Plus measures, do not see a need for them, and worry about their costs to user expression. For example, many DMCA Classic OSPs fear the high cost of developing and using filtering systems, which are sufficiently expensive to represent unsustainable costs for all but the largest players. These OSPs expressed particular concern that small players will not be able to compete effectively in the online services market if filtering becomes a *de facto* requirement. And for many, filtering is inimical to the prioritization of free speech and the careful adjudication of fair use and other rights claims on which their user communities are built.

c. Site-blocking

The most controversial proposed shift away from section 512 notice and takedown, in the end, is the effort to circumvent it altogether—to abandon its complex allocation of rights and responsibilities in favor of site blocking at the level of ISPs. Rightsholders' frustration with what may well approach, in 2016 alone, a billion total takedown requests, helps make this case for blocking a "blacklisted" set of sites. As noted above, for compliant sites focused on music and, to some degree, video, rightsholders have made a successful push for voluntary, norm-based adoption of automated takedown, filtering, and other measures that extend beyond DMCA Classic. For non-compliant sites—including a large number of file-sharing and file-locker sites outside the US—site blocking has been the primary and largely unrealized policy goal.

We have little to add to the technical arguments against site blocking, which generally emphasize the technical ineffectiveness of the measures and the potential danger to the Domain Name System ("DNS").[327] Our research, however, does underscore the due process concerns. Mistakes in detecting and noticing infringement are common; even a system that is (legally if not always practically) limited to removal of specific infringements from platforms severely challenges procedural due process values in current practice. Site blocking, in moving well beyond even filtering and staydown to cutting off access to an OSP in its entirety, raises these process questions far more acutely. In general, remedies as strong as site-blocking require much more process before they are granted, usually a strong showing in court and narrow tailoring. Analogous examples might be temporary restraining orders and injunctions.

Such reservations have not prevented the adoption of site blocking in several countries, including Australia, India, Portugal, and the United Kingdom. At present, the United Kingdom is running a natural experiment in site blocking based on a blacklist of—in late 2015—110 file sharing sites.[328] These measures are too recent and possibly too small in scale to draw conclusions about their effectiveness or the risk they pose. But the legal framework for escalation is in place and the list of blocked sites is almost certain to expand.

In the United States, the issue remains highly controversial. Site blocking provisions were a major contributor to the defeat of the Stop Online Piracy Act ("SOPA") in 2011. Mistakes made by the DOJ in a handful of site blocking actions have fueled fear that it is likely to be abused.[329] We also see no consensus about who should be responsible for identifying and blocking sites. In interviews, rightsholders were very reluctant to take on the potential liability associated with this role, and equally wary of "official" government responsibility, which, in the US, triggers associations with censorship.[330]

Our studies strongly suggest that any such measures should be accompanied by strong statutory due process, a central role for courts, and meaningful liability for mistakes and

---

[327] *See, e.g.*, INTERNET SOC'Y., *supra* note 174; Vixie, *supra* note 174.

[328] Ernesto, *UK Blocking More Than 100 Pirate Sites After New Court Order*, TORRENTFREAK (Mar. 24, 2015), https://torrentfreak.com/uk-blocking-more-than-100-pirate-sites-after-new-court-order-150324/.

[329] *See, e.g.*, Declan McCullagh, *DHS Abruptly Abandons Copyright Seizure of Hip-Hop Blog*, CNET (Dec. 8, 2011, 11:14 AM), http://www.cnet.com/news/dhs-abruptly-abandons-copyright-seizure-of-hip-hop-blog/.

[330] On public attitudes toward government and private sector roles filtering and site blocking, see KARAGANIS & RENKEMA, *supra* note 176.

abuse consistent with the greater potential for suppression of expression. As we have seen, a liability regime that runs in only one direction produces mistakes and over-enforcement.

2. Notice and Takedown as Competition Policy

One of the important goals of section 512 was to prevent copyright liability from becoming a significant barrier to entry in the development of online services. And by all accounts it succeeded. Most of the OSPs in our sample viewed the boom in innovation in services as inseparable from section 512 safe harbor protection. If OSPs are pressured to adopt DMCA Plus measures, however, there is some evidence that this could unravel.

The costs of compliance with new enforcement demands are clearly a worry for OSPs with limited resources, especially insofar as these costs are cumulative. The development of workflow, triage, and takedown systems for automated notices is expensive and does not obviate the need for human review teams. The development or licensing of filtering systems is expensive and does not replace automated notice workflows and triage. Because not all rightsholders participate in filtering system catalogs, OSPs have to maintain parallel processes. Because each system requires human support, staff costs grow. In practice, these cumulative demands create competitive advantage for well-resourced OSPs that can bear the costs of building and maintaining multiple systems.

We are potentially in a moment of change, in which some OSPs have become powerful incumbents, equipped to satisfy rightsholders' concerns, but at the potential cost of moving norms or rules to a place where smaller OSPs cannot follow. The concern is that the safe harbor's availability could become limited to those with resources for DMCA Plus measures, in turn raising barriers to entry and limiting the robust competition that gave rise the vibrant landscape of online services that currently exists. Content ID and other filtering systems, from this perspective, are not a sign of war between rightsholders and Internet companies, but a sign of accommodation between dominant incumbents. This concern is heightened by the fact that many OSPs are outside of the spaces where large-scale infringement problems arise, yet shifting requirements could still negatively affect them.

The consolidation of services into large integrated platforms creates related concerns. The DCMA does not require notices sent to one service to affect access to others owned by the same company, such as advertising and payment services not encompassed by the DMCA. Yet some multi-service providers have begun to connect noticing systems in ways that amplify the power of these third-party accusations. The SoundLocker case described how notices sent to Google Search triggered cancelation of a Google Adsense account—a primary source of revenue for many OSPs. In that case, the linkages between services led to a compliance maze, in which the targeted OSP lost core business services while trying to respond of rounds of notices sent between third parties. In such cases, "compliance" could mean whatever the central platform provider requires.

In Study 1, we briefly focused specifically on Google, reflecting the fact that both rightsholder and OSP respondents tended to describe it as central to notice and takedown's evolution and as, to some degree, affecting others' actions. Accordingly, its evolving approach to notice and takedown is a good example of how a dominant incumbent's decisions can influence copyright practice. Content ID's content filtering and monetization features suggest a possible solution for rightsholders reeling from the disruption of classical distribution models. But ContentID's dominance, and the cost barriers to creating competing systems, also give Google a strongly advantageous position for bargaining with content providers. Its filtering and monetization features also raise well-documented concerns about who decides questions of fair use and restriction of expression, as well as concerns from follow-on creators about control and fair compensation.[331] One example of this tension surfaced when the diverse multi-channel networks—which had operated largely independently on top of YouTube and exercised their own decisions about copyright complaints—were pulled into the Content ID system in late 2013, generating a wave of Content ID matches and setting off a judicially untested argument about the fair use status of video game replays and who can monetize them.[332] Providers are on novel terrain with many of these issues. Yet as large OSPs become platforms for a wide range of activity, notice compliance systems affect all the communities that rely on these platforms to reach an audience, from the paradigmatic YouTube uploader to the growing array of businesses and services that depend on those platforms.

D. RECOMMENDATIONS FOR POLICY AND PRACTICE UPDATES

Our research suggests a number of policy and practice interventions to address problems with notice and takedown, solidify some of its better features, and expand and disseminate best practices that OSPs and rightsholders have developed. As our research was limited to stakeholders' experience with the notice and takedown framework under United States law, our recommendations focus on improvements to that regime. As such, broader reform proposals that operate outside of section 512's framework are not addressed here.[333] In formulating these, we have given special attention to helping those whose problems became most apparent in our research: targets, smaller senders, and less-well-resourced OSPs.

Our work suggests that improving notice and takedown requires addressing some quite different problems. First are mistakes and overbroad takedown caused by automated systems. In this case, rightsholders generally are sophisticated and knowledgeable of copyright.[334] It is the coarseness of automated decisions, and the overwhelming scale, that create challenges. Second are human mistakes, and in some cases, misuse or abuse of the takedown system. In this case, senders are less likely to have sophisticated knowledge of copyright law and notice and takedown.

Statutory reform can help address some of these problems, especially where the statute has created an imbalance. Notably, the statute's bias toward takedown, and the weakness of its remedies for targets, can be improved. Many of our suggestions, however, are practice-based. Our Study 1 respondents—both OSPs and rightsholders—have developed a wealth of experience applying notice and takedown and making changes with shifts in technology and business models. Their experience informs education and best practice recommendations.

---

[331] See, for example, the Future of Music Coalition's analysis of how independent music artists are affected by ContentID and various other fingerprinting and monetization systems. Griffin Davis & Kevin Erickson, *Vimeo Introduces Audio Fingerprinting*, FUTURE OF MUSIC COAL. (May 27, 2014), https://futureofmusic.org/blog/2014/05/27/vimeo-introduces-audio-fingerprinting. While FMC sees benefits, it also criticizes YouTube for focusing on large copyright holders. *Id.*

[332] Google responded by introducing a distinction that conveys different rights and responsibilities—"managed" channels retain responsibility for copyright enforcement and received preferential monetization; "affiliates" operate fully under Content ID. For a useful summary of the dispute around 'Let's Play' game videos and the attendant fair use issues, see Sebastian Mejia, *Fair Play: Copyright Issues and Fair Use in YouTube "Let's Plays" and Videogame Livestreams* (Working Paper, 2013), http://ssrn.com/abstract=2368615.

[333] For an example of an alternative framework for addressing intermediary liability that falls outside of section 512's framework, *see Manila Principles on Intermediary Liability*, MANILA PRINCIPLES, https://www.manilaprinciples.org/ (last visited Feb. 5, 2016).

[334] We note that REO agents may not all be as knowledgeable as the rightsholders themselves; this is an open research question.

We conclude with a set of "anti-recommendations"—potential changes that, based on our research, are likely to cause more harm than good. These focus on preserving the benefits of notice and takedown and meeting section 512's pro-competition goals in the face of expanding automation. Currently, the problems that come with automation, while large in magnitude, are limited to a subset of senders, OSPs, and targets. However, these problems could easily spill over into the rest of the notice and takedown system. Policy and practice reforms should not hasten this outcome.

1. Statutory Reforms

Our recommendations for statutory reforms are relatively modest, and often follow the suggestions of others. Our research suggests that the basic legal rules underlying notice and takedown remain constitutive of both OSPs' and rightsholders' approaches to copyright enforcement. They have been used by parties, and applied by courts, for nearly two decades; it would be counterproductive to disturb them unduly. Changes should not reduce less-well-resourced OSPs' ability to rely on the safe harbor, and any change should take into account the interests of targets, small- and medium-sized copyright holders, and small- and medium-sized OSPs. Importantly, any changes should not disturb the core safe harbor, and should preserve its pro-competition effects.

Our research was also clear, however, that notice and takedown needs better mechanisms for ensuring that an infringement is actually likely before material comes down and stays down, and better due process mechanisms for targets. Some technical fixes would also be beneficial.

a. Mistake and Abuse

One of the most troubling findings in our research was the high number of questionable notices we observed in Studies 2 and 3. This finding, coupled with Study 1 respondents' broad agreement that targets almost never use the counter notice process, and similar findings in all other all empirical research to date,[335] indicates that better methods of preventing and remedying mistaken and abusive takedown demands should be a high priority for reform. Other commentators have come to a similar conclusion; our recommendations include some of their suggestions.

Both OSPs and senders can implement best practices to alleviate some of these harms; we describe these ideas below. In terms of statutory reform, however, addressing mistake and bad faith must focus on the *sender's* good faith and prudence, coupled with better mechanisms for targets. Currently, the standards for sending a notice are low—so long as "knowing, material misrepresent[ations]" are avoided, there is no liability for sending a bad notice. In Study 1, we found broad agreement amongst OSPs that the limited legal incentives to avoid sending mistaken or abusive notices notably contribute to problematic takedowns. Yet takedown is a powerful remedy for a cheap price that has a profound effect on the target; senders should prepare appropriate notices and be prepared to back up their claims. OSPs are not in a good position to adjudicate claims, as they will never know the full context of the complaint (the dispute reflected in Ella Miller's Study 3 notices provides a good example of this). We recommend:

---

[335] *See* Seng, *supra* note 5; Urban & Quilter, *Efficient Process*, *supra* note 5, at 688; Urban & Quilter, Undue Process, *supra* note 100.

- **Improving the quality of notice claims.** We recommend, following Public Knowledge's suggestion, to harmonize section 512(c)(3)(A)(vi) with section 512(g)(3)(c) so that takedown notice senders, like counter notice senders, must declare under penalty of perjury that they have a good faith belief that the substantive claims in a takedown notice are accurate.[336] Copyright claimants—who would bear the burden of showing infringement in court—should have to stand by their substantive claims just as targets must stand by their assertion of non-infringement. This change could encourage copyright holders to consider the validity of their complaints prior to sending, in line with the recent decision in *Lenz v. Universal Music Corp*,[337] and incentivize the improvement of automated infringement detection systems.[338]

- **Allowing immediate putback in response to a valid counter notice.** Section 512(g)(2)'s ten-day waiting period before material goes back up in response to a valid counter notice should be repealed. This has been suggested in various forms by a number of commentators.[339] The ten-day waiting period is routinely criticized for jeopardizing expression, especially time-sensitive expression. Given the very small number of counter notices received by OSPs and the high social cost of censoring expression, any costs related to this change would be far outweighed by the benefit of fixing this problem.[340]

- **Making it more feasible for targets (or OSPs) to recover for harms caused by illegitimate notices.** We recommend following Public Knowledge's suggestion to change section 512(f)'s "knowing, material misrepresentation" standard to disallow "reckless" misrepresentations. This would make it more feasible—though still challenging—to recover against a takedown sender making a bogus claim.[341] This reform is even more important in light of the decisions in the *Rossi v. Motion Picture Association of America, Inc.* and *Lenz* cases, which allow copyright holders to rely on subjective beliefs of the accuracy of their claims.[342]

  Practically, this change would improve the cost-recovery calculus for an OSP or target considering a lawsuit by addressing the fact that the current evidentiary burden of proof is too high for OSPs and targets to feel confident that they will prevail. An important

---

336 *See, e.g.*, CHEN, DURKEE, FRIEND, & URBAN, *supra* note 320, at 14-15.

337 Lenz v. Universal Music Corp., 2016 U.S. App. LEXIS 5025, at *16 (9th Cir. Mar. 17, 2016) (holding that fair use is "authorized by law" and a copyright holder must consider the existence of fair use before sending a takedown notification under § 512(c)).

338 We note that requiring that senders declare good-faith accuracy under penalty of perjury would not preclude automated sending, but would likely create incentives to improve the accuracy of automated processes. Standards for the development and deployment of automated system could be established that meet a reasonable standard for declaring that the substantive claims in the notices they generate are accurate.

339 *See, e.g.*, CHEN, DURKEE, FRIEND, & URBAN, *supra* note 320, at 14-15; Urban & Quilter, *Efficient Process*, *supra* note 5, at 688; Urban & Quilter, Undue Process, *supra* note 100, at 4; Letter from Joan Claybrook & Paul Alan Levy, President & Attorney, Public Citizen, to Senators John McCain & Barack Obama (Oct. 16, 2008), https://www.citizen.org/documents/DMCALetter.pdf [hereinafter Public Citizen Letter].

340 While we hope that the suggested reforms will encourage more counter notices, it is highly unlikely that they would encourage more in situations where copyright infringement is clear. Bogus counter notices sent by overseas pirates are few, and easy to interpret and dismiss. Encouraging appropriate counter notices, however, would simply have the effect of either resolving the dispute—in the case of a clear-cut non-infringement—or sending the parties to court, which is the appropriate venue for truly contested claims.

341 *See* CHEN, DURKEE, FRIEND, & URBAN, *supra* note 320, at 11-12.

342 Rossi v. Motion Picture Ass'n of Am., Inc., 391 F.3d 1000, 1003-4 (9th Cir. 2004); Lenz v. Universal Music Corp., 2016 U.S. App. LEXIS 5025, at *17 (9th Cir. Mar. 17, 2016).

effect would be to give OSPs that are "flooded" by truly abusive notices a better chance to recover against the abusive sender. OSPs in Study 1 universally agreed that section 512(f) currently fails to provide them with meaningful protection from this type of abusive behavior.[343] Giving OSPs a better mechanism for recovery could help weed out some of the abusive uses of automated systems, while leaving their benefits intact for legitimate claims.

- **Reforming the Copyright Act's statutory damages provisions to provide meaningful recovery for violations of Section 512(f).** Currently, section 512(f) only allows recovery of "any damages, including costs and attorneys' fees" by anyone injured by a misrepresentation. Commentators have noted that this provides limited cover for targets or OSPs harmed by wrongful takedown, as actual damages are difficult to define in this context. Recommendations include allowing targets to recover punitive damages[344] or statutory damages.[345] We recommend allowing recovery of limited statutory damages, taking care to design them to be "fair, reasonable, and proportionate to the harm."[346] Done well, such a framework could better balance senders' and targets' rights. (Copyright holders, of course, have statutory damages available them under section 504(c). We suggest similar changes to these remedies below.)

- **Adopting some of the IPTF's suggested statutory damages reforms.** The IPTF recently suggested broader reforms to the Copyright Act's statutory damages provisions in its 2016 *White Paper on Remixes, First Sale, and Statutory Damages*.[347] Some of these reforms would further reduce abuse and assist OSPs in overcoming their current bias toward takedown. As the IPTF's suggestions also include reforms that extend beyond addressing mistake and abuse, we discuss them in more detail next.

b. Statutory Damages Reform

In Study 1, OSPs explained that they fear very high potential liability for their users' infringements, and that this motivates conservative behavior. The high statutory damages currently available under US copyright law are a main source of their fear. OSPs are acutely aware that statutory damages, which range from $200 to $150,000 per work infringed, multiplied across the thousands or millions of works that users might infringe, could easily sink a company. This drives them to be biased toward takedown and creates market entry and competition worries.

Some sensible reforms, such as guidelines for courts suggested by Samuelson and Wheatland, and more recently, statutory reforms suggested by the IPTF, could alleviate these issues while leaving in place robust remedies for copyright holders.[348] Rebalancing statutory damages could also help address abusive takedowns by giving OSPs more confidence to reject notices.

---

[343] In Study 1, OSPs with which we spoke about this issue explained that the costs of trying to use section 512(f) were too high to provide any meaningful relief from abusive senders. *See* Section III.C.3. Changing the standard to "reckless" would still provide a relatively high bar, but should help OSPs recover from truly abusive senders who flood them with "trash" notices.

[344] Urban & Quilter, *Efficient Process*, *supra* note 5, at 690.

[345] *See, e.g.*, CHEN, DURKEE, FRIEND, & URBAN, *supra* note 320, at 13; Public Citizen Letter, *supra* note 339.

[346] CHEN, DURKEE, FRIEND, & URBAN, *supra* note 320, at 21-22.

[347] DEP'T OF COMMERCE INTERNET POLICY TASKFORCE, *supra* note 206.

[348] For additional recommendations for what courts should and should not do when awarding statutory damages under the current Copyright Act regime, see SAMUELSON & WHEATLAND, *supra* note 314, at 501-09.

It was the potential for statutory damages that led one respondent to describe every decision not to take down material in response to a notice as "bet[ting] the company" (see Section III.B.2.a). More broadly, statutory damages reform could buttress the safe harbor's pro-competition effects by giving start-ups and smaller OSPs, which cannot absorb as much risk, more leeway in designing business models.

We recommend adopting the sensible reforms proposed by the IPTF in its white paper,[349] along with some modest extensions:

- **Giving courts discretion to depart from the "per infringed work" calculation in cases of non-willful secondary liability for online services.** Currently, the statute directs courts to calculate statutory damages by applying a set amount to each work infringed.[350] This "per infringed work" provision did not anticipate the numbers of works that users can place onto large online platforms or the absurd results—requests of billions or even trillions of dollars in damages—that can ensue. Fear of such staggering liability costs can cause OSPs to make overly conservative decisions regarding notice and takedown compliance. This is especially true when OSPs make decisions that could affect their eligibility for the safe harbor in the first place, because the loss of the safe harbor could result in liability for a potentially large number of infringed works. For example, one Study 1 respondent views the only safe reading of "repeat infringer policy" as requiring the suspension of posting rights for users when it receives a second notice targeting that user's content (see Section III.C.5).

  Giving courts discretion to depart from the "per infringed work" calculus, where appropriate, would provide a more hospitable environment for investment and innovation in online services while still giving courts the ability to redress and deter infringements. Under the IPTF's proposed model, courts would still employ the "per infringed work" standard in cases of willful infringement and in cases where they deem it otherwise appropriate.

- **Specifying factors in the Copyright Act that courts must consider when assessing statutory damages.** Currently, section 504 provides scant guidance to decision makers determining where in the broad range of potential statutory damages an award should fall, increasing uncertainty for everyone involved. The IPTF recommended addressing this by incorporating in the statute a list of factors for courts and juries to consider when determining the amount of statutory damages to award.[351] Overall, adopting this list of factors would provide greater predictability to rightsholders and OSPs and would help deter abusive threats to targets. In addition, some of the factors proposed by the

---

[349] Dep't of Commerce Internet Policy Taskforce, *supra* note 206, at 86-99.

[350] 17 U.S.C. § 504(c).

[351] The IPTF's proposed factors are 1) the plaintiff's revenues lost and the difficulty of proving damages; 2) the defendant's expenses saved, profits reaped, and other benefits from the infringement; 3) the need to deter future infringements; 4) the defendant's financial situation; 5) the value or nature of the work infringed; 6) the circumstances, duration, and scope of the infringement, including whether it was commercial in nature; 7) In cases involving infringement of multiple works, whether the total sum of damages, taking into account the number of works infringed and the number of awards made, is commensurate with the overall harm caused by the infringement; 8) the defendant's state of mind, including whether the defendant was a willful or innocent infringer; and 9) In the case of a willful infringement, whether it is appropriate to punish the defendant and if so, the amount of damages that would result in an appropriate punishment. Dep't of Commerce Internet Policy Taskforce, *supra* note 206, at 87-88.

IPTF—such as a factor that weighs the defendant's financial resources—would support newcomers, thus promoting competition and supporting smaller and new OSPs.

- **Increasing the availability of statutory damages reductions in cases of innocent infringement.** Currently, the Copyright Act allows courts to reduce statutory damages for innocent infringements if the defendant "had no reason to believe" the act was infringing.[352] Statutory damages can also be remitted, but only for a limited set of nonprofit defendants and only in limited situations.[353] The IPTF recommends changing the statute so that the existence of a copyright notice is no longer a bar to the assertion of an innocent infringer defense. This is sensible. For example, this change would increase the likelihood that OSPs might reject a takedown request targeting material that is likely protected by fair use but bears a copyright notice.

  Based on our research, we also recommend further changes that the IPTF declined to include on the record before it. A substantial number of problematic takedown requests in Study 2 and Study 3 invoked claims outside copyright or raised issues other than fair use. Clarifying that high statutory damages are less likely to be available to a plaintiff in such cases could help targets more confidently send counter notices and OSPs more confidently reject problematic notices. Accordingly, we recommend revising section 504(c)(2) to give courts discretion to reduce or remit statutory damages for any defendant who reasonably believed that the challenged use was not infringing. This change would shift the standard for innocent infringement enough to allow courts discretion in cases of fair use or other exceptions—not only cases where there was "no reason to believe" at all that there could be infringement—and would allow remitter, if appropriate, for any defendant.

c. Transparency

The opacity surrounding notice and takedown should be addressed more fully than current OSP "transparency report" efforts—as valuable as they are—can provide. Takedown is a strong remedy, with no public oversight in all but the tiny proportion of cases that are disputed and make it into court. The challenge is facilitating transparency without causing unwarranted harms to rightsholders, OSPs, or targets. In interviews, rightsholders worried about revealing enforcement tactics that could be exploited by pirates, and OSPs worried about attracting undue attention from either rightsholders or targets for publishing notices. Targets—who are the subject only of allegations, not proven claims—have privacy interests at stake.

Taking all of this into account, we recommend considering Public Knowledge's proposal that notice and counter notice senders submit notices to a centralized repository, available to be searched and analyzed. In order to prevent undue costs for rightsholders and targets, we recommend:

- A simple interface that would allow for one-click sending both to the relevant OSP and the repository. It should be easy to use for less-sophisticated rightsholders, and should not be burdensome for large rightsholders who need to submit notices in bulk.

---

[352] 17 U.S.C. § 504(c)(2).

[353] *Id.*

- Requiring the repository to have the capacity for machine submission and analysis through open Application Programming Interfaces ("APIs").

- Redacting any individual notice targets' names and contact information from the publicly available side of the repository.

- At least temporarily redacting the names and contact information from the publicly available side of the repository of individuals who send notices or counter notices for themselves. While it is true that senders choose to send notices, and would have to reveal their names when filing a suit, our research shows that individuals do not always have the sophisticated understanding of copyright law that section 512 presumes, and we worry that public naming could chill sending.

- Policies that allow researchers to access the notices in the repository, beyond the redacted publicly available versions.

Who should operate such a repository is an open question. The Copyright Office, if it is given sufficient resources, is an attractive option to provide integrated notice and takedown support. It already maintains the required list of DMCA agents, and could couple the notice repository with educational information to help senders and targets use the system appropriately.

This recommendation does not replace the efforts of OSPs (and rightsholders, should they choose to do so) to publish transparency reports. Indeed, we also recommend that OSPs and rightsholders that do not currently publish transparency reports on their copyright notice and takedown activity consider doing so. Transparency reports—especially when they give an idea of the disposition of the complaint on the part of the OSP—provide a valuable public service.

d. Technical Fixes

Section 512 is regularly criticized for its technical complexity and, at times, ambiguity. Others have suggested a variety of technical fixes.[354] Our interviews highlighted a technical issue—whether all types of OSPs are protected for putback in response to a valid counter notice—that has otherwise not been discussed, and that should be corrected.

Section 512's structural design is based on categorizing different types of OSPs and assigning them somewhat different responsibilities, while also including provisions that appear to apply to all OSPs. In doing so, Congress left unclear the important question of when a non-hosting OSP must accept a counter notice, and when it is protected for "putback" in response to a counter notice. In practical terms, this creates a disincentive for search providers to accept or act on counter notices.

The issue is highly technical, and almost certainly unintentional. Section 512(g)(1) states that OSPs are protected from:

---

[354] Public Knowledge advocates clarifying the safe harbors to apply to all rights that are copyright subsidiaries or otherwise closely related, specifically, user uploads of "bootlegs," which fall under a specific provision in addition to copyright law; material that violates the anti-circumvention provisions; and false copyright management information. CHEN, DURKEE, FRIEND, & URBAN, *supra* note 320, at 5-6. In the same paper, Public Knowledge suggests simplifying the standards for designating the DMCA agents required for safe harbor protection.

"any person for any claim based on the service provider's good faith disabling of access to, or removal of, material or activity claimed to be infringing or based on facts or circumstances from which infringing activity is apparent, regardless of whether the material or activity is ultimately determined to be infringing."

This protects all OSPs from suit for takedown. Section 512(g)(2) then creates an exception to the 512(g)(1) safe harbor that would allow targets to sue an OSP that does not notify the target of the takedown and respond to any valid counter notice the target sends. Section 512(g)(4) then clarifies that service providers complying with 512(g)(2)—that is, responding to a counter notice with a putback—are protected from copyright infringement liability if the material turns out to be infringing after all. Taken together, 512(g)(2) and (4) are meant to balance the strong bias toward takedown created by 512(g)(1).

The issue arises because service providers are protected under 512(g)(4) for putback if they comply with 512(g)(2), but 512(g)(2) appears only to apply to hosts: the language of 512(g)(2) applies only to "material residing at the direction of a subscriber of the service provider on a system or network controlled or operated by or for the service provider..."

This leaves the status of search providers murky. Most people assume that search providers are also obligated to accept counter notices and are protected for putback, and this is debatably true. But it is not clear. For it to be the case, links to search results would have to be considered material that resides "at the direction of the subscriber." The muddiness of the issue is even more apparent because it is widely recognized that search providers, which do not have a direct subscriber relationship with their users, are not obligated to notify users of removal under 512(g)(2)(A).

The easiest technical fix is to revise Section 512(g)(2) to apply to any "material that is removed, or to which access is disabled by the service provider, pursuant to a [takedown] notice..." This would make clear that search providers must respond to counter notices and that they are protected against liability for putback. However, it would also obligate search engines to take "reasonable steps promptly to notify the subscriber that it has removed or disabled access to the material." To some degree, this change is attractive: we were troubled by the number of questionable takedown notices we found in Studies 2 and 3, and even more by the fact that many targets likely did not know about the notices because they were sent to search services. Requiring search providers to take "reasonable steps" to notify targets would alleviate that. But it is unclear what those "reasonable steps" would be; without any service relationship with targets, search providers have little to go on. Practically, this solution would create a large burden on search OSPs without much evidence that targets would actually be found.

Given this, the best fix would make clear that search providers must accept and respond to counter notices, and that they are protected for putback. Section 512(g) should be revised to clarify that sections 512(g)(2)(B), 512(g)(2)(C), and 512(g)(4) apply to all providers.

2. Best Practices, Information-Sharing, and Education

The diversity of practice among both OSPs and rightsholders, and their lack of insight into each others' practices, were notable observations in Study 1. In this situation, organized guidance in the form of best practices and educational materials could supplement statutory reform by sharing information and increasing the quality and accuracy of notice and takedown practice. For a variety of issues these methods may be more effective at improving practice than one-size-fits-all statutory changes. Best practices and educational materials could be developed privately or with agency guidance, but any process initiated to develop them must take into account the interests of the full range of stakeholders, including targets.

We derived a number of the following recommendations from our Study 1 respondents' descriptions of their knowledge, procedures, and practices. After analyzing all of our data, we arrived at additional suggestions. Finally, the Department of Commerce DMCA Multi-stakeholder Forum's list of "good, bad, and situational practices" for notice and takedown reflect the variety of practice, and we largely agree with the treatment of the practices outlined in the document.[355]

a. Effective Enforcement: Avoiding Mistakes, and Preventing Abuse

While refining the statutory standards and remedies can help address mistakes and abuse over time, good practices are essential to limiting operational problems.

i. Automation Problems

For sophisticated rightsholders, whose mistakes appear to be driven by automation and scale, improved algorithms and better human review practices are needed to limit problematic notices and focus efforts most accurately on large-scale infringement. In Study 1, rightsholder respondents described a number of practices intended to avoid overtargeting during automated campaigns. OSPs filled in some of the gaps by describing overtargeting they had observed, and our quantitative studies prompted further suggestions. Recommended best practices for rightsholders and REOs using automated systems include:

- **Developing automated infringement detection and notice-sending policies and practices that better avoid inadvertently targeting noninfringing content. Recommended measures some rightsholders use include:**
  - Focusing automated efforts on "rogue" sites, which are identified through human review, rather than sending automated notices in response to any machine-flagged infringement. (See Section III.B.2.a.)
    - Avoiding inadvertently targeting noninfringing content requires limiting the definition of "rogue" site to those that truly exist to provide a platform for infringing file sharing, streaming, and download. One rightsholder, for example, described first contacting sites to see if "you can find them and negotiate with them"—which indicates an intention to operate legitimately—or if they "operate in the shadows."
    - Rightsholders' lists of identified "rogue" sites should be regularly reviewed and updated in order to avoid flooding OSPs with notices that target defunct sites, or target sites that no longer exhibit "rogue" practices. (See below and Section IV.B.2.a.)
  - Going beyond simple filename or URL matching to matching entire files through fingerprinting (see Section III.B.2.a), and then employing human spot-checks of algorithmic decision making.

- **Exercising care in choosing agents, and policing their work.**
    - Holding regular "bake-offs" with REOs to help ensure both that rightsholders are getting the best service and that REOs' methods avoid overbroad takedown. (See Section IV.B.2.a.)

[355] *See supra* note 326.

- **Working with OSPs.** Working with OSPs who receive a large number of notices to identify ways to streamline enforcement but avoid overbroad takedown. (See Section III.D.)

The rightsholders who engage in many of the above practices lower the risk that large-scale takedown practice will damage expression rights. However, the relatively high number of mistakes that still appeared in Study 2, and the features of those mistakes, strongly suggest further best practices:

- **Employing systematic algorithmic and human cross-checks to avoid questionable notices:**
  - Develop searching techniques that include "flags" for possible mistaken identity, fair use, permissioned use, and the like, and route flagged alleged infringements for human review prior to dispatching a takedown notice.
  - Regardless, apply human review to randomized samples of machine-generated infringement results prior to dispatching takedown notices. Use what is learned from this technique to better tailor search methods, "flags," and further sampling methods.
  - Periodically subject automated methods to testing and review by expert third parties and revise as needed.
  - Take specific measures to avoid sending duplicate requests. This was a repeated complaint from OSPs.
  - Relatedly, avoid sending to defunct sites. This was also a repeated complaint from OSPs, and something we observed in Study 2. (See Section IV.B.2.a.) We note that our observation involved a relatively small number of REOs and other senders, all of whom were targeting obvious file-sharing sites. Other senders managed to avoid defunct sites. It seems that this is a fixable problem.

- **Developing technical and human methods for better identifying allegedly infringing materials** and identifying them in notices. Study 2 identified significant problems with the identification of the allegedly infringing material targeted for removal. For example, many notices provided a URL that led to a search results page or aggregator page that included multiple works, making identifying the allegedly infringing material problematic. (See Section IV.B.2.c.ii.) Others linked to pages where the allegedly infringing material was embedded on the linked page. (See Section IV.B.2.f.) Automated techniques should be refined to identify the direct address of the allegedly infringing material in question, not a page on which it resides.

- **Identifying copyright owners of the works identified in notices.** Nearly 37 percent of Study 2 notices failed to specify the copyright owner of the allegedly infringed work; typically these notices were from trade associations that listed "member companies" as the copyright owners of the identified work. Senders should specify the copyright owner of the AIW so that OSP recipients and targets can fully evaluate the claims in the notice and respond.

- **Employing policies and guidelines for enforcement agents and enforcement staff to help them identify fair use and relevant tolerated uses.**

- **Ensuring that REOs, if used, follow best practice identification and notice-sending policies.** This includes:
    - Developing policies and practices that include the technical methods and human spot checks described above; requiring REOs to implement them in service contracts; and periodically reviewing REOs' practices to verify that they are in compliance with the policies.
    - Educating REOs, like in-house enforcement staff, about the rightsholders' policies regarding tolerated uses. (See Section III.C.3 for a discussion of the problem of REOs taking actions with which rightsholders would not agree.)

The last practice is crucial: REOs or other agents sent nearly 92% of the notices in Study 2. Ensuring that they follow good practice is key to improving the accuracy of large-scale takedown.

For their part, OSPs that use automated measures described a variety of crosschecks and triaging systems that "escalated" some notices for human review. Good practices they described included:

- **Developing automated methods to weed out obviously flawed notices (such as requests to remove home page URLs).** (See Section III.D.1.b.)

- **Conducting human spot-checks of automated notices to refine the automated methods that flag questionable notices.**

- **Escalating questionable notices for human review.** (See Section III.D.a.)

- **Conducting automated review and human spot-checks of notices submitted by "trusted" senders and removing senders that have high rejection rates from their trusted sender programs.** (See Section III.D.1.b.)

- **Ensuring that trusted sender systems still require formal DMCA notices, identification of the infringed work, and notifying the target of the takedown request.** (See Section III.D.1.b.)

- **Tailoring the takedown of hash-matched files to limit overbroad takedown.** (See Section III.D.1.c.i.)

Not all OSPs described following the above practices. We recommend them as best practices for any OSP using automated systems. To them we add:

- **Where the scale of notices is too great for human review of every notice, subjecting random samples of notices to human review** (as also suggested for rightsholders).

All of these practices are recommended in addition to following the "Good General Practices for Notice Senders" and avoiding the "Bad General Practices for Notice Senders" identified in the IPTF's "List of Good, Bad, and Situational Practices."[356] While implementing these practices may increase costs somewhat—as they require more human review and more rigorous development and maintenance of automated systems—the benefit of limiting the

[356] *Supra* note 326.

harmful effects of automated decision-making is high, and taking the recommended steps should not unduly affect rightsholders' ability to engage in large-scale infringement policing.

ii. Human Error and Smaller Senders

Highly automated systems' mismatch with the fact-based nature of copyright infringement poses a unique set of problems, but sending notices by hand, based on human review, unfortunately does not guarantee quality.

Both Study 1 and Study 3 make clear that some senders—especially those who are less experienced with copyright law or the notice and takedown process—need better information in order to avoid sending problematic notices. The struggles these notice senders face appear to reflect the complexity and esoteric nature of copyright law, and the fact that differences between copyright, trademark, privacy, defamation, harassment, and other troublesome speech may be subtle and unclear to non-experts.[357] Unlike the sophisticated rightsholders who rely on and exercise control over automated systems, these senders are more likely to benefit from guidance at the point of sending a notice. Lack of knowledge may also be behind some cases of misuse or abuse by these senders—it is not always clear when abuse is intentional and when it is based on misunderstanding the copyright system. This creates a role for OSPs and, potentially, for institutional actors.

While stressing that they are not in a position to adjudicate copyright disputes, many of the OSPs with whom we spoke in Study 1 do reject notices for a range of reasons. Indeed, as noted above, some OSPs' published rejection rates are as high as 60–80%.[358] Like rightsholders, OSPs described a range of beneficial practices. These included:

- **Using human review, and questioning or rejecting notices in appropriate circumstances.** While OSPs cannot engage in general adjudication, the current limitations of the counter notice process mean that some OSP review and adjudication is an important check on abusive takedowns.

- **Engaging in additional scrutiny of notices that display certain "flags," such as a one-time or new sender, a sender that is an apparent competitor of the target, or unusual or grammatically error-ridden text.** These notices may well be legitimate; however, OSPs noted that in their experience, such "flags" can indicate a notice that is more likely to exhibit problems. (See Section III.C.3.)

- **Providing senders with educational materials and clear guidance about appropriate copyright takedown requests.** OSPs that created educational materials to explain subject matter appropriate to copyright complaints, web forms that helped senders formulate their complaints properly and provide sufficient information to comply with section 512, educational materials for senders, and similar information, found that their efforts increased the quality of takedown requests. These materials should not be off-putting to senders, but should cover copyright limitations—such as limitations on subject matter, and fair use—as well as copyright rights. Educational materials need

---

[357] Beyond the examples in Studies 1 and 3, Urban and Quilter highlighted some of the correspondence between The Planet and notice senders who appeared to misapprehend copyright law and the DMCA's requirements. *See* Urban & Quilter, Undue Process, *supra* note 100, at 21-22.

[358] Reddit, Reddit Transparency Report, *supra* note 12 (reporting a 62% rejection rate in 2014); *Requests for Content Alteration & Takedown*, Wikimedia, *supra* note 12 (reporting a 86% rejection rate between January and June 2015).

to be available at the point of filling out the notice and sending it for those who would otherwise not encounter the information because they are not generally focused on copyright issues. (See Section III.C.3.)

It was less common for OSPs to discuss measures to help targets. Because of the current imbalance in section 512's protections for targets compared to senders, OSPs were understandably wary of encouraging counter notices. Still, some offer educational information and clear counter notice mechanisms.[359] The best practice is:

- **Providing targets with educational materials and an easy-to-use counter notice function.** Providing information about copyright protections and limitations in an unbiased, unthreatening manner—ideally, the same information provided to senders would serve both purposes—can help.

This may be especially helpful for more specialized OSPs that serve as platforms for specific communities with local ideas of appropriate use that may not fully correspond to copyright law.

For OSPs with fewer resources or more general platforms, providing access to a centralized information source could help. We next turn to potential agency, multi-stakeholder, or other institutional efforts to provide educational materials and tools for senders and targets.

iii. Increasing Access to Notice and Takedown and Counter Noticing

Smaller copyright holders, like smaller OSPs, are far less likely than large, well-resourced industry players to have their interests reflected in policy debates. And targets of notices may be the most underserved group in the notice and takedown ecosystem, as stakeholders with actual experience as targets are absent from policy discussions. As noted, the counter notice mechanism has largely failed, and OSPs do not feel comfortable offering guidance to targets who are trying to understand whether a takedown claim is proper. Shared investment would help improve notice and takedown for both of these groups.

Investing in information resources that all senders can access before deciding to send notices, and that targets can access before responding to them, is an important step. OSPs could link to these resources and ask senders to review them before submitting a notice or counter notice. The resources could be created and hosted by the Copyright Office, the Internet Policy Task Force, another governmental body, or an outside institution like a law school clinic or academic center. They should include:

- Information about copyright law and its limitations, and how notice and takedown reflects these rules.

- The differences between copyright and commonly confused issues like trademark and defamation.

[359] *See, e.g.*, *DMCA Counter Notice*, AUTOMATTIC, https://automattic.com/dmca-counter-notice/ (last visited Feb. 5, 2016); *Guide to Submitting a DMCA Counter Notice*, GITHUB, https://help.github.com/articles/guide-to-submitting-a-dmca-counter-notice/ (last visited Feb. 5, 2016); *When Should I File a Counter-notice*, TWITTER, https://support.twitter.com/articles/15795#9 (last visited Feb. 5, 2016).

- How to prepare and send a takedown notice, including:
    - Considering fair use or other limitations;[360]
    - Identifying the copyrighted work sufficiently for the OSP to review the notice;
    - Identifying the alleged infringement sufficiently for the OSP to remove it;
    - Understanding the target's counter notice options and when filing a lawsuit would be necessary; and
    - Understanding the potential ramifications of sending an improper notice.

- And for targets, similar information to help them understand whether a counter notice is appropriate and how to send it, including:
    - Considering the copyright claim in the notice;
    - Understanding copyright limitations and the scope of the right to send a counter notice;
    - Understanding section 512's requirement that counter notice senders accept US federal court jurisdiction; and
    - Understanding the potential ramifications of sending an improper counter notice.

Though these materials would be helpful regardless, making the counter notice process truly usable requires the statutory rebalancing suggested in Section V.D.1.

In Congressional testimony,[361] representatives of independent artists and other smaller senders also criticized a lack of meaningful access to the more sophisticated enforcement methods (including automation, REO contracts, and monetization strategies) available to larger copyright holders. Although we did not speak with small senders directly, we note that the competition issues that arise with content filtering and monetization issues (see Section III.E.) also affect independent artists, smaller labels, and other individual creators. Further information-sharing and research efforts would also be beneficial:

- Exploring ways to make monetization models more available to independent artists, smaller labels, other individual creators, and follow-on users.[362]

- Exploring how to make automated tools to search for potential infringements more available to independent artists, smaller labels, and other individual creators.[363] Crucially, these tools must follow the best practices outlined above in order to avoid exacerbating the issues we observed with mistake and abuse. In general, any expansion of automated systems should occur in combination with the liability-balancing measures suggested in Section V.D.1 in order to ensure that mistaken or abusive notices are minimized.

---

[360] *See* Lenz v. Universal Music Corp., 2016 U.S. App. LEXIS 5025, at *16 (9th Cir. Mar. 17, 2016) (holding that fair use is "authorized by law" and a copyright holder must consider the existence of fair use before sending a takedown notification under § 512(c)).

[361] *See, e.g., Statement from: Copyright Alliance CEO Sandra Aistars, supra* note 153 (stating that independent authors lack the resources of corporate copyright owners and cannot "dream of the robust enforcement programs that larger companies can afford").

[362] For example, advocacy groups supporting independent musicians have noted that since YouTube's Content ID was introduced, independent artists and smaller labels have expressed frustration with the process of getting their works included in the database for matching. *See Vimeo Introduces Audio Fingerprinting*, FUTURE OF MUSIC COAL., https://futureofmusic.org/blog/2014/05/27/vimeo-introduces-audio-fingerprinting (last visited Feb. 5, 2016).

[363] Audible Magic, the lead fingerprinting database for music, allows independent artists and smaller labels to upload five free files to the database. *Id.* Additional uploads are available for a fee. *Id.*

- Encouraging a market for REOs that follow best practices and will be available to independent artists, small labels, and other individual creators at a reasonable price.[364]

3. Changes to Avoid

Our research also provided three key insights into changes *not* to make to the notice and takedown system. First, the relatively high number of problematic notices we observed in Study 2 counsels against expanding automated practices without much better controls against mistake and abuse. Second, DMCA Classic OSPs appear to make up a substantial portion of the online ecosystem, and they are very sensitive to the costs automated measures would impose on them. Further, many DMCA Classic OSPs rarely or never encounter large-scale infringement issues, weighting the cost-benefit analysis against automated measures. Third, the numbers of problematic notices we observed in our quantitative studies show that targets' expression interests remain a crucial issue for notice and takedown policy.

Given these insights, DMCA Plus measures should remain entirely voluntary. Policy measures should not:

- Require filtering, staydown, or automated systems more generally.

- Change the knowledge standard for OSPs so that DMCA Plus systems are practically required for OSPs to obtain and retain the safe harbor.

- Expand takedown requirements to tertiary "Para DMCA" providers. These providers have little or no insight into underlying copyright disputes, yet including them can create the type of serious negative consequences for legitimate providers illustrated by the SoundLocker case study.

We are sympathetic to the difficulties some rightsholders face in policing their copyrights in the digital environment, and in some cases DMCA Plus measures have helped manage infringement. At the same time, it is senders who are in the best position to improve accuracy and avoid the worst mistakes. More importantly, our research strongly suggests that the tradeoffs that would come with requiring these types of measures—rather than using them in targeted, voluntary situations—are simply too high. The risk of exacerbating due process issues for targets and limiting market entry and competition for OSPs combine to create a potent danger of undermining the current use and future development of online expression platforms.

E. FUTURE AVENUES FOR RESEARCH AND FACT-FINDING

As noted throughout, our research was limited by a lack of information in some areas. With regard to formulating good policy, the interests of targets and smaller copyright senders are especially understudied. A number of topics that would be beneficial to explore include:

[364] *See supra* note 308 for small senders' perspectives on enforcement.

- Smaller copyright holders' knowledge of copyright, experience with notice and takedown, and both enforcement and distribution needs.[365] We are very grateful to the copyright holders we interviewed, who provided valuable insight into a major group of takedown notice senders. However, our respondents were all large copyright holders with valuable properties and significant resources for enforcement. We expect that most copyright holders will have far fewer enforcement resources, may depend more on third-party OSPs for distribution, may have less sophisticated strategies for copyright exploitation, and may exhibit other differences than affect their interests in the notice and takedown system.

- Targets' knowledge of copyright and experience with notice and takedown. This has proven difficult to examine as targets are currently known only to the sender and OSP. Further, it is difficult for some researchers to study targets, due to ethical concerns regarding human subjects (because targets may be infringers whose practices are revealed in research). But understanding targets' reasoning, how they approach copyright complaints, and their knowledge of copyright law, is key to knowing whether the "due process lite" mechanisms provided in notice and takedown regimes are effective.

- Experimental work that "tests" notice and takedown systems, especially automated systems, to see if practices aimed at avoiding overbroad takedown are effective.

- Economic analyses of the costs created by different DMCA Auto and DMCA Plus measures, how effective they are for rightsholders, and if and how they affect competition and market entry for OSPs.

- Research into the REO market for information on how REO firms work, the mechanisms they use, best practices they follow, costs for copyright holders, and how much of the takedown universe they occupy. Our REO interview material was helpful, but thin, as we were not able to interview many REOs. A wider picture would be beneficial.

---

[365] We think there is a substantial need here. For small senders' perspectives on enforcement, *see supra* note 308. The Future of Music Coalition recently pointed out that independent music artists may differ in their needs and interests from larger copyright holders, and called for further information:

> [B]efore entering a policy battle to amend or modify the existing requirements, there should be a consideration of how changes might impact independent creators, content publishers, developers and users. It is entirely possible that DMCA S. 512 creates conditions for a suppression of market rates for music licensing. We want streaming services to succeed and for artists to be paid more as adoption increases; if data can demonstrate that safe harbors are impacting growth, a policy response is warranted. But first we need the data. It is also possible that DMCA-enabled services are among the few viable options for bringing a product forwards in an environment of incredible consolidation among content companies.

FMC Filing in the Office of the Intellectual Property Enforcement Coordinator of the US 2016 Joint Strategic Plan, FUTURE OF MUSIC COAL., http://www.futureofmusic.org/filing/fmc-filing-office-intellectual-property-enforcement-coordinator-us-2016-joint-strategic-plan (last visited Feb. 5, 2016).

And as noted above, our Study 3 suggested that less-sophisticated senders could use help with substantive copyright issues, and our Study 1 respondents agreed that these senders were most likely to send problematic notices. *See supra* Sections III, IV.

- Research into specific areas of active enforcement. We spoke with a range of rightsholders, and found that taking time to delve more deeply into the needs of different sectors (for example, movies or publishing) would likely produce interesting results. Research into the adult entertainment industry's use of notice and takedown, especially, would be helpful, as copyright holders in this sector are some of the most active users of notice and takedown, and there is little information about their practices other than the material we obtained from OSPs in this study.

- Further quantitative and qualitative empirical work into how notice and takedown operates on the ground, especially by sector, and as technology and business models continue to change. As important as Google is to the notice and takedown ecosystem, our qualitative work shows that there is much more to it than Google reaches. Further substantive work on notices sent to OSPs other than Google is crucial to gain a fuller picture than we could generate.

VI. CONCLUSION

Eighteen years on, notice and takedown as enshrined in section 512 of the DMCA is the core mediator of Internet copyright disputes. The overall picture that emerged from our research suggests that the fundamental compromise behind notice and takedown remains crucial, even though Internet services and online infringement have both evolved significantly in the years since it was passed. The safe harbor's availability remains a considerable factor in OSPs' ability to enter the market, survive, and develop, and the takedown process is a linchpin of enforcement.

Notice and takedown, however, is not wholly satisfactory to any party in the chain. Our research supported the view that, as practiced today, it exhibits some serious defects. Of particular concern, limited due process for targets pairs with, in our samples, a relatively high proportion of problematic takedown requests.

Automation proved to be a flashpoint in our studies. While its growing use by some important rightsholders and OSPs represents a major shift, it is also far removed from the practice and experience of most individual OSPs. OSPs' widely varying capacity for implementing expensive automated measures, and the due process issues that come with them, strongly suggest that imposing these measures through statutory reforms or practice norms that circle back into legal requirements through litigation would be ill-advised. The latter is perhaps the most concerning. If too many well-resourced incumbent OSPs and rightsholders agree on expensive, *ex ante* measures that are practically available only to some and that undermine the copyright limitations that protect expression, ensuing norms could both filter lawful expression out of major platforms and radically alter the ability for new or niche OSPs to provide competing platforms.

It may seem quixotic to try to improve notice and takedown when the most prominent voices in the debate focus on ever-escalating wars between robots and pirates, abusive takedowns, and offshore infringement. But its enduring importance to online speech platforms and copyright holders alike show the value in trying. Relatively modest reforms to better balance the interests of targets and senders, coupled with the best of the practices described to us in our research, would help address some the due process challenges created by notice and takedown, while leaving in place the inexpensive, rapid remedy it provides to rightsholders.

APPENDIX A: ADDITIONAL DESCRIPTIVE STATISTICS ON ALLEGEDLY INFRINGED WORKS AND ALLEGEDLY INFRINGING MATERIAL

STUDY 2: SIX-MONTH LUMEN DATASET

Audiovisual content accounted for 44% of the allegedly infringed works ("AIW") identified in the coded takedown requests (predominantly requests sent by or on behalf of principals in the adult entertainment or movie/television industry). The AIW was identified as audio in 42% of requests (predominantly from the music industry), and as visual in 22.1% of requests (predominantly from the adult entertainment industry). Software accounted for just under one in ten—8.5%—of AIWs, and written material for 3.1%. (These numbers add up to more than 100% because some works fall into more than one category.)[366] *See Fig. 16*, below.

TYPES OF MEDIA ALLEGED TO BE INFRINGED IN TAKEDOWN REQUESTS

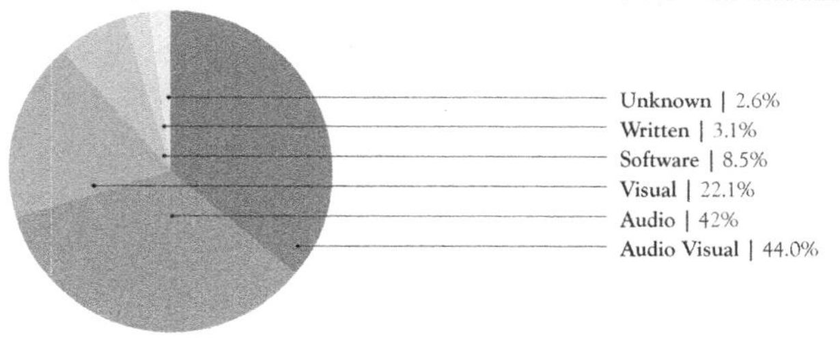

Figure 16: Types of Media Represented in Takedown Requests[367]

We then reviewed the complained-of links and examined the media targeted for takedown—the allegedly infringing material ("AIM")—where available.[368] This allowed us to independently identify what type of media was targeted for takedown. *See Fig. 17*, below.

[366] For example, the adult entertainment industry often identified the allegedly infringed work as both video and images, and, in the absence of any further identifying information in the notice, coders tagged the allegedly infringed work as both audiovisual and visual.

[367] Numbers add to more than 100% because some AIWs fall into more than one category.

[368] As noted in the methods and discussed in more detail in Section IV.A.1, the AIM was not always available for review.

TYPE OF MEDIA: ALLEGEDLY INFRINGED WORKS COMPARED TO ALLEGEDLY INFRINGING MATERIAL

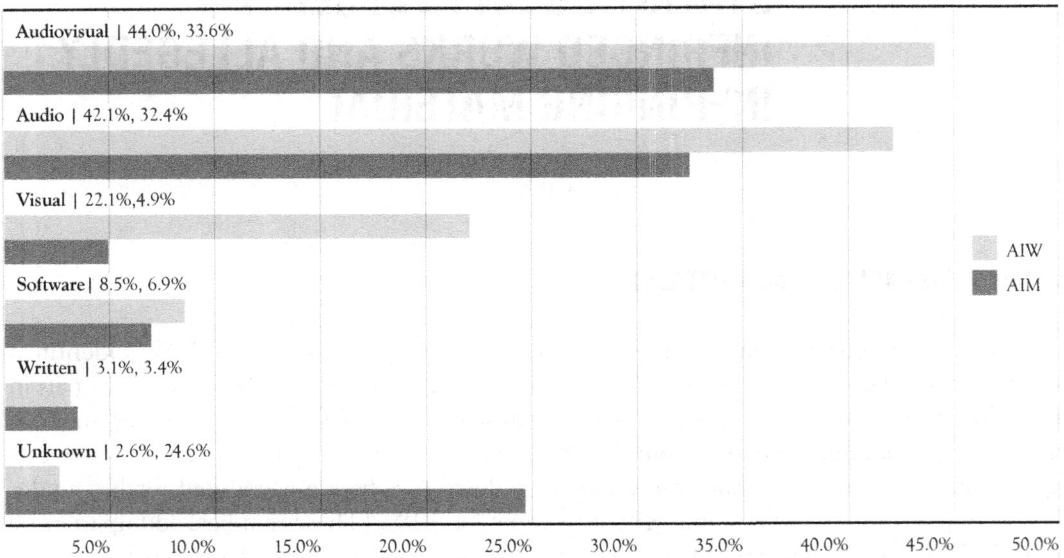

Figure 17: Type of Media: Allegedly Infringed Works Compared to Allegedly Infringing Material

As might be expected, the types of media targeted for takedown as AIMs generally track the types of media senders allege are infringed. 33.6% of the requests targeted audiovisual content, 32.4% targeted audio content, 4.9% targeted visual content, 6.9% targeted software, and 3.4% targeted written material. There is, however, some variance, often based on how senders described the AIW in their notices and which of several layered copyrights they might be enforcing. Senders representing the adult entertainment industry, for example, often claim both "video and images" as the AIW, while the AIM file, once reviewed, is clearly only one or the other. In the opposite vein, senders in the music industry often identify the AIW with an artist and song title only. This leads the AIW to be classified as audio, while the AIM might be an audiovisual music video. Finally, a sizeable difference follows from lack of available information about some AIMs. As described above, coders were unable to classify nearly a quarter (24.6%) of AIMs because the content was unavailable.

STUDY 3: SIX-MONTH LUMEN DATASET—GOOGLE IMAGE SEARCH TRANCHE

Unsurprisingly for an image search service, most senders claim infringement of pictorial and graphic works. Almost all of the requests from senders other than Ella Miller—97.9%, in fact—relate to pictorial and graphic works as at least one component. (16.9% of the non-Miller requests claim infringement of written content; most of these are also categorized as pictorial and graphic works. These typically are cases where a copyrighted image is part of a work that also includes textual elements—such as a greeting card). A negligible number of the allegedly infringed works in the non-Miller requests—well below the margin of error—were identified as audio-visual (.2%) or software (.1%).[369] *See Fig. 18*, below.

[369] Because content could be classified as more than one category, the total percentage is over 100%. For example, some requests identified the allegedly infringed work as both an image and its accompanying text.

TYPES OF MEDIA ALLEGED TO BE INFRINGED IN TAKEDOWN REQUESTS (GOOGLE IMAGE SEARCH SAMPLE)

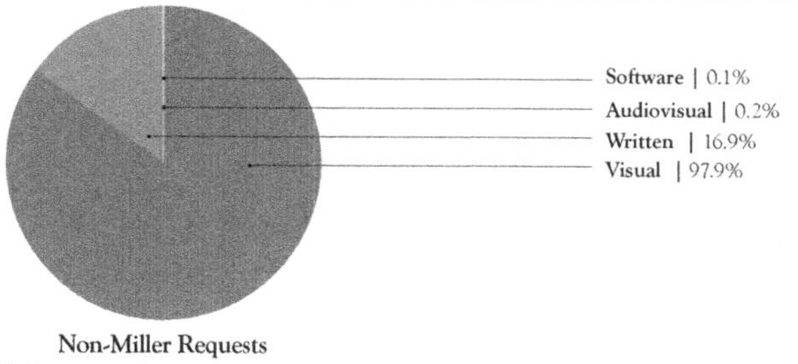

Figure 18: Types of Media Alleged to Be Infringed in Takedown Requests (Google Image Search Sample; Non-Miller Requests)

Unsurprisingly, the type of allegedly infringing media targeted for takedown largely tracked the media type of the identified work alleged to be infringed. The majority of the non-Miller requests targeted allegedly infringing visual content (86.3%). 13.8% of the non-Miller requests targeted written content, a small number targeted audiovisual content (3.3%) and a single request targeted software (.1%). *See Fig. 19*, below.

TYPES OF ALLEGEDLY INFRINGING MATERIAL (GOOGLE IMAGE SEARCH SAMPLE)

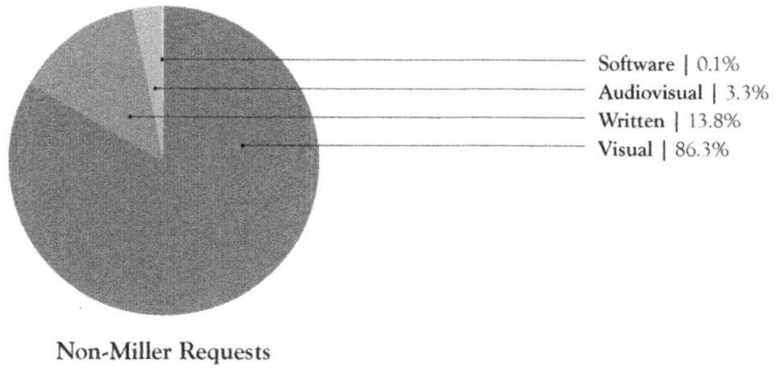

Figure 19: Types of Allegedly Infringing Material (Google Image Search Sample; Non-Miller Requests)

APPENDIX B: STUDY 1 DATA DOCUMENTATION

A. METHODS

Study 1 provides a descriptive account of the takedown ecosystem. The primary data are original, drawn from a detailed survey and a set of semi-structured interviews. We focused first on 29 intermediary online service provider ("OSP")[370] respondents, including platform, connectivity, and search providers. To these we added interviews with six major notice senders (including rightsholders, rightsholder groups, and rights enforcement agents). These original data were combined with publicly available information and secondary literature and reporting.

We began with OSPs because their position in the center of the notice and takedown process means that they receive and act on notices and counter notices from a wide range of participants. This gives their staffs a broader view of the process than any one group of senders or targets. We supplemented with large senders in order to get a picture of how this important constituency experiences notice and takedown.

We reached the survey and interview group using a multi-layered selective sampling approach. We initially identified OSPs to contact using a list of the top-100 most-visited websites in the United States,[371] and then distributed the survey to these service providers' designated DMCA agents using the list of designated agents on the US Copyright Office website. Using the top-100 websites both limited researcher bias and ensured that initial respondents were likely to have sufficient traffic to engage with the notice and takedown process. We then employed a modified snowball approach. Survey recipients were encouraged to forward the survey to other potential respondents. The widely broadcast survey led to interviews, and interview respondents recommended other respondents. In order to maximize our chances of covering typical cases, we also approached OSPs who clearly represented a category or size of service not represented in our initial outreach. With regard to rightsholders, we focused on supplementary interviews with large rightsholders with strong interests in the operation

[370] Respondents included video and music hosting services, search providers, file storage services, e-commerce sites, web hosts, connectivity services, and other user-generated content sites (e.g. blogs, social media, or other specialized platforms). Eleven of our OSP respondents' services were oriented towards individual users (e.g. user-generated content or entertainment services) and 18 were oriented towards both enterprises and individuals (e.g. cloud software and storage, productivity tools). 15 were small companies with fewer than 100 employees, six were medium companies with 100-499 employees, and eight were large companies with 500 or more employees.

[371] Alexa Top Sites in the United States, http://www.alexa.com/topsites/countries/US (last visited Feb. 5, 2016). Of the 100 emails sent to DMCA agents using the contact information provided on the list of designated agents on the US Copyright website, 18 of these emails failed to deliver (generating automated server error response).

of notice and takedown.[372] Some of the initial set of service providers and rightsholders then forwarded our request on to other potential participants. We took care to solicit participation from a wide range of OSP stakeholders who were diverse in size, type of service, and position within the ecosystem and from a cross-section of major rightsholders. The result is not a statistically representative sample of the OSP or rightsholder communities, but it is a broadly inclusive one with more than one representative of nearly all of the major categories of service, as well as important differentiators within those categories.

We were unable to include the targets of notices or categories of notice senders other than large senders. Although OSPs were able to provide some information about those participants, their absence limits the reach of our study. For more information about our methods and their limitations, see pages 26-28 of the report.

B. OSP QUESTIONS AND INTERVIEW METHODS

Survey. Below are the questions that comprised our survey of OSPs. Responses to these questions, together with information from semi-structured interviews with OSPs and senders, formed the basis of our Study 1 findings.

Interviews. Interviews with OSPs were semi-structured. Initial OSP interview questions were based on these survey questions. However, in interviews we often used open-ended forms of the questions that were appropriate for interviewing but not useful in the survey format. For example, the interview format allowed us to ask open-ended questions like, "Please tell us about your takedown process," and then follow up with more specific questions to elicit detail. Interviews also included organic follow-up questions suggested by the discussion. Additionally, a small number of new topics, not on the initial survey, arose in interviews. For example, in interviews, some OSPs described providing senders backend access to and direct takedown privileges on their systems. Where relevant, we brought these topics up in subsequent interviews.

1. What is your name and position?
2. What is the name of your company?
3. What type of services does your company provide? (e.g., connectivity, hosting, software as a service, etc.)
4. Roughly how many employees does your company have?
5. Where are notices received? (i.e. location of the office)
6. Approximately when did your company start receiving copyright takedown notices? (year and month if possible)
7. How many separate notices (which each may include multiple links or claims):
   Do you receive on average per week?
   Did you receive in total in all of 2013?
8. Do you know or can estimate the number of separate notices (which each may include multiple links or claims) received in:
   2012
   2011
   2010

[372] These included trade associations, large organizational rightsholders, and rights enforcement agents. They represented a range of the industry sectors that most use notice and takedown: games, movies, music, software, and the like.

9. Do you know (or can you estimate) the total number of URLs or alleged infringements cited in theses notices:
   2013
   2012
   2011
   2010
   Sorry, we can't determine this easily.
10. How many of these notices were sent under a regime other than the U.S. DMCA (e.g., South Africa's similar regime):
   2013
   2012
   2011
   2010
   Sorry, we can't determine this easily.
11. Within these non-US notice submissions, do you know (or can you estimate) the number of URLs or alleged infringements submitted in each of these notices?
12. Which country or countries' laws are included in these notice submissions? (e.g., Canada, China, South Africa)
13. Do you use a web form or a dedicated email form to receive takedown notices?
14. If yes, when did you introduce it?
15. If yes, has it had any obvious impact on the number of notices you receive?
16. If yes, has it had any impact on the type of notices you receive? (e.g., more or fewer copyright notices? More automated notices? More junk notices?)
17. About how much person-power do you employ to review notices? (e.g., 2.5 full-time positions worth)
18. How many lawyer-positions does this number include? (e.g., 1 full-time position)
19. Is your process supplemented by automated filtering and/or notice processing of some kind?
20. If yes, can you describe this system?
21. If yes, does this system ever escalate to human review?
22. If yes, please describe when escalation occurs:
23. Do you have a "Trusted Sender" program or another method for differentiating certain notice senders?
24. If yes, can you describe it briefly?
25. If yes, what benefits or challenges does this system create?
26. How many counter notices did you receive in:
   2013
   2012
   2011
27. About what percentage of takedown results result in:
   content takedown
   putback
   Has this changed over the past few years?
28. What are the most common bases for rejecting a takedown request?
29. Can you tell us about your policies regarding users who are the subject of repeated takedown request? Do you employ a "three strikes," "six strikes" or similar policy to deal with such situations? Do you terminate accounts for repeat takedown requests? If so, under what conditions?
30. For ISPs participating in the Copyright Alert System: How many notices have you received in the last year and do you evaluate them differently than traditional DMCA notices? If so, how any with what overall results?

31. Have upstream providers ever notified you that they received a takedown demand based on user content on your site?
32. If yes, have they ever notified you of or taken any action against you (like pulling your site down temporarily)?
33. Do you publicly report on your notice and takedown activities?
34. If no, why not?
35. What are the main benefits of the notice and takedown process under Section 512?
36. What are the main challenges or problems with the notice and takedown process under Section 512?
37. Are there other issues around notice and takedown that you would like to emphasize or flag to us for research?
38. Would you be willing to discuss sharing your company's notices for research purposes?
39. Are there questions in this survey that you could not answer? Or do not wish to answer?
40. If you're willing to follow up, please leave us your email address.

C. SENDER INTERVIEW QUESTIONS AND METHODS

Below are questions representative of those used to guide interviews with senders.

As with OSPs, these interviews were semi-structured. Initial questions were designed to cover similar ground as those used with OSPs, but from a sender perspective. For example, while OSP respondents were asked about how many notices they *receive*, senders were asked about how many notices they *send*. Similarly, while OSPs were asked to describe their takedown process, senders were asked to describe their primary strategies and methods for identifying infringing material and sending takedown notices. Other questions were equally suitable for both OSPs and senders. For example, both were asked about the benefits and challenges of notice and takedown. As with OSPs, we often asked senders open-ended questions, then followed up with more directed questions when needed to elicit detail. And as with OSPs, interviews also included organic follow-up questions suggested by the discussion.

1. What are your primary methods for addressing online infringement? What role does notice and takedown play?
2. Please tell us about your notice and takedown strategy and practice.
3. What are your primary methods for identifying infringing material?
4. Around how many notices did your company or association send in the past year? In the year before?
5. Who were the primary recipients in the past year? In the year before?
6. Do you categorize the recipients, or prioritize them in some fashion? Or have other systematic approaches to how you send notices?
7. Do you use automated systems to identify infringing content? If so, can you describe them?
8. Do you use human review for some/all notices? Why or why not?
9. What role does search play in your takedown process? Do you typically send notices to search services?
10. Do you participate in any trusted partner relationships with OSPs, like the Google "Trusted Partner" program? If yes, how do these processes work for you? How are they better or worse than earlier arrangements?
11. Do you use rights enforcement services to assist with notice and takedown? Please tell us about how you use these services and your experience with them.
12. Do you send copyright notices under laws other than the DMCA? Roughly how many or what percentage?
13. Are there practices on the recipient side (e.g., form-based submission) that make your job easier? Harder?

14. Do you follow up to evaluate the success/failure of a notice? If so, how?
15. What is your experience with counter-notices? How many of these have you encountered in the past year? How do you address them?
16. How do you address potential fair uses or permissioned or tolerated uses?
17. Are there "good" and "bad" actors in the cyberlocker, filesharing, etc. worlds? If yes, what are the differences? Are there "pirate" sites that are responsive to notices?
18. Have your methods evolved since you started? If so, how?
19. Are your practices the norm in the notice-sending world? If not, how are they different?
20. What are the main benefits of the notice and takedown process under Section 512?
21. What are the main challenges or problems with the notice and takedown process under Section 512?
22. Does the notice and takedown process "work" in your view? What would you change, if anything?
23. Are there other issues around notice and takedown that you would like to emphasize or flag to us for research?
24. Would you be willing to discuss sharing your company's notices for research purposes?
25. [*If interviewee is a rights enforcement agent*] Can you relate how and when you got into this business? What kinds of companies were/are your clients?
26. Are there other rightsholders or agents that you think would provide a useful perspective on these issues (and be willing to talk to us)?
27. Have we missed anything important?

APPENDIX C: STUDY 2 AND STUDY 3 DATA DOCUMENTATION

A. METHODS

1. Dataset and Coding

For more information on Study 2 and Study 3 data and methods, see pages 78–82 and 98 of the report.

We requested from Lumen all unredacted takedown requests in its repository for the period between May 1, 2013 and October 31, 2013. The Lumen team used the following query to pull the data we requested:

SELECT * from tNoticePriv where date >= DATE_SUB(now(), interval 6 MONTH)
This returned 288,675 separate takedown notices, containing a total of 108,331,663 takedown requests.

We adopted the individual takedown request—rather than the notice body, which might have many requests included within it—as the primary unit of analysis. More than one takedown request can be made in an individual notice, and each takedown notice body may contain dozens or hundreds of requests. From the rightsholder's perspective, each request is a claim to remove one allegedly infringing item. From an OSP's perspective, each is a claim that must be acted upon. An accurate analysis thus requires reviewing individual takedown requests. Because of this, researchers should be aware that more than one request may have been coded from a single Notice ID (the "notice_noticeid" variable).

Data were imported into a customized database for review and coding. At the point of importation, machine-readable information was extracted and auto-populated into the relevant database fields. A randomization function randomly selected and presented takedown requests to reviewers for coding.[373]

Notice characteristics were then coded by reviewers—the authors and a team of law student research assistants under the supervision of the authors—as described further on page 81 of the report. The resulting variables are described below in Section B.4.

Further information about the sampling used for Study 2 and Study 3 can be found in Sections B.1 and C.1 below.

[373] The randomization function supported both sampling across the entire dataset and building randomized "tranches" of more targeted subsets while maintaining overall randomness. In this paper, we used two samples: one randomized sample taken from across the entire six-month set (Study 2) and one randomized sample taken from the Google Image Search notices (which were isolated from the entire six-month set) (Study 3).

The coded and redacted samples and Stata 13 files are available to anyone who agrees to follow certain research ethics guidelines. Access to these data, descriptive documentation of the variables used in Study 2 and Study 3, and more information about the requirements, are available at http://takedownproject.org/data-release. We note, however, that "link rot" limits the ability to analyze the data over time, as material in question is taken down or moves. We were unable to fully analyze just over a quarter of the notices for this reason (see discussion pages 79-80) and we expect that this issue will increase with time. Researchers should be aware of this problem, fully disclose its extent in any reporting based on the data, and clearly describe their methodology for managing it.

2. Data Processing

Data were prepared for statistical analysis in a five-step process. First, the raw data files were exported from the database into CSV files and then into the Stata 13 statistical package; second, sample sizes were confirmed; third, fields from the coding database were matched with the raw data to check for missing data; fourth, standard missing data analysis procedures were employed to properly identify missing data and any potential patterns therein and, where appropriate, replace missing information;[374] and fifth, uni- and bivariate statistical analyses were conducted on the cleaned data to examine frequency distribution and correlative patterns. The majority of the analysis was purely descriptive.

Composite Variables

Composite variables based on section arrays were created using STATA's *egen* command, which takes a series of related measures and computes a score based on responses. This results in a summed measure, which was then used to indicate the absence of presence of a given state. Using this, a new dichotomous measure was created, coded as 1 if that state was present or 0 if not. All requests included in the 1 category were those whose score on the composite measure was greater than or equal to 1. The *egen* composite measures were also used to confirm requests that were coded as more than one item in an array.

Please note: if a request was coded as more than one item, it was counted only once for purposes of that composite measure. A less conservative approach that counts each item appearing in a notice separately could be appropriate, but we preferred the more conservative approach of counting the notices that exhibited *at least* one characteristic in the composite measure.

[374] Of primary concern among these selections was to distinguish between legitimately missing data and instances in which the missing data indicated a condition simply did not apply. Each section of questions was examined for skip and missing patterns (to check whether missing data points randomly distributed across the takedown notices and not concentrated among specific types of notices, senders, or reviewers) and then matched to the data in the Stata files. Analysis of skip and missing patterns revealed that the only non-randomly missing information pertained to mutually exclusive question arrays (see below for more details). There were no patterns identified among types of notices, business sectors, or content material. For each section of questions there are two forms of variables: stand alones, whose answers are not dependent or contingent on previous or next questions, and arrays, whose answers are dependent or contingent on previous or next questions. For stand alones, all missing cells remained as such and were not recoded. For example, if the address of the attorney associated with the takedown selection was not included, it remained missing. For arrays, all missing cells were carefully cross-matched with the section of questions they came from. Those that were not applicable rather than missing were recoded appropriately. For example, for each selection there are arrays of questions pertaining to the size of the business that posted the selection. Business size is a mutually exclusive category (a subject cannot be both a small business and a large business). In these cases, selections that were coded as a specific business size were recoded (i.e. the data was relabeled as "0", rather than missing) for all other business sizes. Once these arrays were determined, one set of Stata syntax was written and then edited using find and replace to properly treat each section of the data.

Please see Section C.2 below for discussion of some additional processing that was required for the Study 3 data.

Redaction of Individual Principals' Names

We redacted the names of individuals where they appeared in the coded sample. Principals were categorized as individuals for redaction purposes if the "notice_sender_principal" field included an individual's name and the "notice_sender_lawfirm" field did not include a business name (typically, this field was either blank or repeated the individual's name).

Where the principal was identified as an individual, the principal's name was redacted from the "notice_sender_principal", "notice_sender_lawfirm", and "notice_sender_attorney" fields. The individual's name was replaced with "[REDACTED [#]]." Each redacted individual was assigned a number so that researchers can identify when a single sender was responsible for more than one request in the coded sample. In two cases, instead of redacting the name, we changed the individuals' names to match the pseudonyms we used in the paper (Ella Miller and Anatoli Ivanov).

Where the individual's name appeared in the description of the allegedly infringed work (the "work_display_name" variable), the name of the author of the allegedly infringed work (the "work_coding_author" variable), the name of the copyright holder of the allegedly infringed work ("work_coding_copyright_holder_sta" variable), in the text in a link provided to identify the allegedly infringing material (the "link_display_name" variable), or in the name of the user, site, or service targeted by the request (the "analysis_target_name" variable), the individual's name was redacted from the text and replaced with "[REDACTED]."

B. STUDY 2

1. Sampling

Where the source data were machine-readable, we were able to run queries across the entire six-month dataset of 108,331,663 takedown requests, obviating the need to sample.[375] We were able to generate the following results in Study 2 from the entire six-month set:

- the total percentage of takedown requests submitted using an automated Google notice submission form (page 83 of the report);
- the total percentage of requests that came from members of Google's Trusted Copyright Removal Program (page 83 of the report);
- the total percentage of takedown requests that were requests to Google Web Search (page 83 of the report);
- Table 2: Google Services Represented in Lumen (page 84 of the report); and
- Figure 6: Top Senders Targeting File Sharing Sites Dead More Than 8 Months (page 90 of the report).

More detailed analysis required sampling. The remaining Study 2 findings are based on a random sample of 1827 requests drawn from the six-month dataset. As the full dataset contains 108,331,663 takedown requests, this sample provided a margin of error of +/- 2.29 at a 95% confidence interval, and +/-3.02 at a 99% confidence interval.

[375] The full, uncoded dataset can be requested from Lumen using the method we described in Section A.1.

2. Variables

Individual variables reported on in Study 2 are described in the following tables. For purposes of analysis, composite variables comprising multiple individual variables were created as described in Section A.2 above. These are further described in the table in Section B.5 below.

3. Errata

Total Potentially Problematic Requests

In an oversight, we neglected to report the total number of potentially problematic requests observed in Study 2 in the original published report—instead, we only reported the separate counts for "fundamentally flawed" requests and for "questionable" requests. We have corrected this in the revised paper and documented the methodology for the counts in the tables below. In short:

- Total fundamentally flawed requests (any request labeled "No Match"): 4.2% (77)
- Total "questionable" requests (any request labeled with any of the SR_que, FU_que, SM_que, or Other_que variables): 28.4% (518)
- Total potentially problematic takedown requests, including both fundamentally flawed requests and "questionable" requests: 31% (566)
    - Note that this is not a simple addition of the "fundamentally flawed" variable total with the "questionable" variable total because a number of requests were tagged with individual variables from both sets. Following our general methodology, we included such requests only once in this count.

Statutory Requirements

The composite variable used for reporting the total number of requests failing to meet the DMCA's statutory due process requirements (see IV.B.2.c on page 92 of the report) did not include an important sub-category: problems with the request's identification of the allegedly infringed work (reported separately in the same section). The revised total is 19.5% (up from 15.4%), from a total count of 357.

Other

We note additional minor errors in the text of the original published report (and corrected in this version):

- Page 82 (and footnote 221, page 81): Study 2's sample size was mistyped as 1826; it should be 1827.
- Figure 2 (page 84): "Rights Enforcement Organizations" should be 49.3% (as correctly stated in the text and depicted in the figure, but mistyped as 43.9% in the figure).
- Page 86 (and Figure 5, pages 87): "File Search" should be 33.9% (reported as 34%).
- Page 90: Figure 6 should be titled "Top 8 Senders Targeting File Sharing Sites Dead More than 8 Months" (not 18 months as originally mistyped).
- Page 93: "Deficiency in sufficiently identify the allegedly infringing material" should be 13.3% (reported as 10.6%). One of the component variables, link_coding_unavailable_navigation, was mistakenly omitted from our original count.
- Page 94: "Deficiency in sufficiently identifying the allegedly infringed work" should be 6% (reported as 4.6%). One of the component variables, work_coding_tag_cannot_complete, was mistakenly omitted from our original count.
- Page 95: "Presented a fair use question" should be 6.6% or "one in fifteen" (reported as 7.3% or "one in fourteen"). See "Study 2: Composite Variables" below for further discussion.
- Page 96: "Anti-circumvention" should be 1.1% (reported as 1%).
- Page 145 (and Figure 16, page 145): Audio AIW should be 42% (reported as 42.1%).
- Page 146 (and Figure 17, page 146): Written AIW should be 3.4% (reported as 3.5%).

4. Study 2: Variable Counts

CATEGORY	VARIABLE(S) USED	COUNT	PERCENT
Figure 1: Sender Role (page 84)			
Agent	notice_coding_sender_role_agent	1677	91.8%
Principal	notice_coding_sender_role_princi	137	7.5%
Unknown	N/A - Remainder of 1827 total	13	.7%
Figure 2: Agent Type (page 84)			
Rights Enforcement Agent	notice_coding_agent_enforcement	900	49.3%
Trade Association	notice_coding_agent_trade_associ	694	38.0%
Other or Unknown	N/A - Remainder of 1677 sent by agents	78	4.3%
Law Firm	notice_coding_agent_law_firm	5	.3%
Figure 3: Industry of Principal (page 85)			
Music	notice_coding_principal_ind_musi	803	44.0%
Adult Entertainment	notice_coding_principal_ind_adul	514	28.1%
Movies/Television	notice_coding_principal_ind_movi	311	17.0%
Software	notice_coding_principal_indsware	137	7.5%
Games	notice_coding_principal_ind_game	98	5.4%
Other	notice_coding_principal_ind_oth	97	5.3%
Books	notice_coding_principal_ind_book	76	4.2%
Web Design	notice_coding_principal_ind_web	7	.4%
Photography	notice_coding_principal_ind_pic	3	.2%
Figure 5: Types of Target Sites (page 87)			
Torrent	analysis_target_role_torrent	655	35.9%
File Search	analysis_target_role_file_search	620	33.9%
Other or Unknown	analysis_target_role_other	232	12.7%
Cyberlocker	analysis_target_role_cyberlocker	180	9.9%
Aggregator	analysis_target_role_aggregator	107	5.9%
Fan Site	analysis_target_role_fan_site	81	4.4%
Video Streaming	analysis_target_role_video	68	3.7%
E-Commerce	analysis_target_role_ecommerce	17	.9%
Social Media	analysis_target_role_social	9	.5%
Personal Website/Blog	analysis_target_role_blog	5	.3%
News	analysis_target_role_news	1	.1%
Section IV.B.2.b. Mistargeting 2: Targeted Material Does Not Match the Allegedly Infringed Work (page 90)			
This variable indicates that the AIM plainly appears to be unrelated to or to copy none of the AIW ("No Match")	link_coding_amount_none	77	4.2%
Section IV.B.2.c. Meeting and Not Meeting the DMCA's Statutory Due Process Requirements (page 92)			
Characteristics that raised questions about whether the request met statutory requirements	SR_que SR_que is a composite variable that combines variables that indicate potential non-compliance with statutory requirements; see "Study 2: Composite Variables" below for a description of this variable.	357	19.5%
Deficiency in sufficiently identify the allegedly infringing material	link_coding_unavailable_aggregat; link_coding_unavailable_search_r; link_coding_unavailable_multiple link_coding_unavailable_navigation	243	13.3%

Deficiency in sufficiently identifying the allegedly infringed work	work_coding_id_link_home or work_coding_id_yes_web_unclear, only if they appear without any of these variables: work_coding_id_yes_single work_coding_id_yes_author work_coding_id_yes_apparent work_coding_id_yes_description work_coding_id_yes_list work_coding_id_no_list work_coding_id_yes_link_list work_coding_id_no_link_list work_coding_id_yes_in_aim work_coding_id_other work_coding_tag_cannot_complete	109	6%
Section IV.B.2.d. One in Fifteen Notices Presented a Fair Use Question (page 95)			
Characteristics that indicate checking for fair use	FU_que FU_que is a composite variable that combines variables that indicate a need to check for fair use; see "Study 2: Composite Variables" below for a description of this variable.	120	6.6%
Section IV.B.2.e. Subject Matter Other than Copyright (page 96)			
Subject matter other than copyright (total)	SM_que SM_que is a composite variable that combines variables that potentially indicate subject matter other than copyright; see "Study 2: Composite Variables" below for a description of this variable.	42	2.3%
Trademark	work_coding_tag_extra_trademark; link_coding_claim_trademark; link_coding_inferred_trademark	23	1.3%
Anti-circumvention	link_coding_tag_extra_key; work_coding_tag_extra_key; link_coding_claim_anti_circumven; link_coding_inferred_anti_circum	20	1.1%
Section IV.B.2.f. Other Issues (page 96): This is a "catch-all," or "miscellaneous," category for items that appeared in relatively small numbers of requests.			
Other issues (incorrectly identify REO as copyright owner, vaguely identifying copyright owner, targeting embedded material, targeting pages that appear to be authorized, targeting competitors)	Other_que; Other_que is a composite variables created to capture a range of potential issues; see "Study 2: Composite Variables" below for descriptions of this variable.	66	3.6%
Total Potentially Problematic			
Total Questionable requests	Que_total Que_total is a composite variable that includes any request marked as SR_que, FU_que, SM_que, or Other_que. See "*Study 2: Composite Variables*" below for descriptions of each of these variables.	518	28.4%
Total Potentially Problematic requests, including both the "questionable" requests and the fundamentally flawed "No Match" requests	Total_PP Total_PP is a composite variable that includes any request marked as SR_que, FU_que, SM_que, Other_que, or link_coding_amount_none. See "*Study 2: Composite Variables*" below for descriptions of each of these variables. (This was not originally reported but is included in the updated text of the report.)	566	31%
Appendix A; Figure 16: Types of Media Alleged to be Infringed in Takedown Requests (page 145) These numbers add up to more than 100% because some content fell into more than one category.			
Audiovisual	work_coding_tag_av	803	44%
Audio	work_coding_tag_audio	768	42%
Visual	work_coding_tag_visual	403	22.1%
Software	work_coding_tag_software	156	8.5%

Written	work_coding_tag_written_work	57	3.1%
Unknown	Remainder that are uncategorized	48	2.6%
Appendix A; Figure 17: Type of Media: Allegedly Infringed Works Compared to Allegedly Infringing Material (page 146) These numbers add up to more than 100% because some content fell into more than one category.			
AIW types as in Fig. 16 above.			
Audiovisual	link_coding_tag_av	614	33.6%
Audio	link_coding_tag_audio	592	32.4%
Visual	link_coding_tag_visual	89	4.9%
Software	link_coding_tag_software	126	6.9%
Written	link_coding_tag_written_work	63	3.4%
Unknown	Remainder that are uncategorized	450	24.6%

5. Study 2: Composite Variables

SR_que **Statutory Requirements** This composite variable includes variables that indicate questions related to the statutory requirements takedown requests must meet. See 17 USC § 512(c)(3)(A)(i-vi). Particularly for items related to identifying the AIW and AIM, these categories are based on how these requirements were usually interpreted by courts at the time the report was published. See report page 18 and accompanying footnotes, and page 94, footnote 277.
notice_coding_sender_role_unconn notice_coding_signature_no notice_coding_sufficient_none notice_coding_good_faith_no notice_coding_accurate_no notice_coding_authorized_no work_coding_id_no notice_coding_signature_cannot_d notice_coding_sig_cant_detunaut notice_coding_sig_cant_det_false notice_coding_sufficient_cannot_ notice_coding_good_faith_cannot_ notice_coding_accurate_cannot_de notice_coding_authorized_cannot_ work_coding_tag_cannot_complete link_coding_stop_no_information link_coding_stop_cannot_find link_coding_unavailable_navigati link_coding_unavailable_multiple link_coding_unavailable_search_r link_coding_unavailable_aggregat work_coding_id_link_home *(only if it appears without:* work_coding_id_yes_single work_coding_id_yes_author work_coding_id_yes_apparent work_coding_id_yes_description work_coding_id_yes_list work_coding_id_no_list work_coding_id_yes_link_list work_coding_id_no_link_list work_coding_id_yes_in_aim work_coding_id_other) work_coding_id_yes_web_unclear *(only if it appears without:* work_coding_id_yes_single work_coding_id_yes_author work_coding_id_yes_apparent work_coding_id_yes_description work_coding_id_yes_list work_coding_id_no_list work_coding_id_yes_link_list work_coding_id_no_link_list work_coding_id_yes_in_aim work_coding_id_other)

FU_que **Fair Use** This composite variable includes variables that indicate characteristics that may weigh favorably toward fair use. We were especially interested in automated decisionmaking. The goal was to identify whether there were takedown requests for which some form of human review would be necessary to check for fair use; it was not to attempt to develop an algorithm that would weigh many characteristics to make a final decision regarding fair use. Accordingly, other characteristics of a flagged request may or may not weigh toward fair use. With a remix, for example, the obvious transformativeness of the use may ultimately be offset by the amount used and/or the effect on the market for the original work. One example of course reserves may ultimately be fair use; another may not. An AIM that uses 50% or less of the work may or may not be a fair use in the end, depending on other factors. And so on. The count thus does not reflect a final determination, but a flag for human review. We selected these variables based on: 1) characteristics that are generally considered to weigh in favor of fair use—for example, a transformative purpose for the use, or a relatively small amount taken for the use—and that 2) reviewers could reasonably ascertain objectively from the information available in the notice (including information reasonably available at the location given by the links provided for the AIW and AIM). For this reason, we did not ask reviewers to decide if a use was, for example, "transformative," but instead asked them to simply identify characteristics of the AIM or its use that strongly suggest transformativeness (e.g., the AIM is a remix). Similarly, we did not ask them to judge whether the AIW was likely to have, at best, a "thin" copyright (making it more likely that a follow-on use falls within fair use), but instead asked them to simply identify AIMs that were rulesets, recipes, and the like. Because our methodology relied on using these more objective characteristics, we likely undercounted the number of works that exhibited characteristics that potentially favored fair use. At the same time, no one characteristic is dispositive of fair use, and the fair use factors must all be considered together for a full analysis—any actual determination would need substantial further review. As noted above in Section B.3, we have corrected this composite variable, which caused a minor change to the count. Due to a miscommunication in building the composite, a small number of variables were originally included that did not fit our criteria. Finally, we received some useful feedback regarding the paper's original mention of musical "covers" in the discussion of this variable (page 95). Our discussion was unclear, and some readers got the impression that we counted being a "cover" as, by itself, a potential fair use characteristic. We did not—indeed, we did not create a variable for covers, at all (itself an oversight). Any request flagged as "fair use questionable" was tagged with at least one of the potential fair use characteristics listed here. The issue with covers arose because coders noted to us that some AIMs *otherwise tagged with a variable included in FU_que also appeared to be covers, or partial covers,* and we considered that sufficiently interesting to note. We regret the (understandable) confusion and have corrected the text of the paper.
work_coding_tag_extra_news work_coding_tag_extra_research_a work_coding_tag_extra_educationa link_coding_tag_write_news_art link_coding_tag_written_newspape link_coding_tag_written_book_sch link_coding_tag_write_jour_schol link_coding_tag_write_jourschart link_coding_tag_written_report link_coding_tag_written_course_r link_coding_tag_written_textbook link_coding_tag_written_journal_ link_coding_tag_written_instruct link_coding_tag_written_recipe link_coding_tag_written_rules link_coding_tag_written_referenc link_coding_tag_written_menu link_coding_tag_written_phrase link_coding_tag_audio_remix link_coding_tag_audio_ringtone link_coding_tag_audio_lecture link_coding_tag_av_documentary link_coding_tag_av_trailer link_coding_tag_av_instructional link_coding_tag_av_remix link_coding_tag_av_fan_remix link_coding_tag_av_compilation link_coding_tag_av_news link_coding_tag_av_lecture link_coding_tag_visual_photo_new link_coding_tag_visual_photo_thu link_coding_tag_extra_news link_coding_tag_extra_research_a

link_coding_tag_extra_research_t link_coding_tag_extra_educationa link_coding_tag_extra_user link_coding_market_personal_use link_coding_market_private_share link_coding_market_curated link_coding_market_context link_coding_market_comments link_coding_market_other_comment link_coding_market_parody link_coding_market_concept link_coding_market_criticize link_coding_amount_barely link_coding_amount_small link_coding_amount_lt25 link_coding_amount_lt50 analysis_target_educational analysis_target_role_news analysis_target_role_educational analysis_target_role_library
SM_que **Subject Matter** This composite variable includes all variables that indicate characteristics of the AIW or AIM that raised questions related to the copyrightability of the AIW and/or the underlying legal claim at issue. These categories are based on caselaw and other information, such as Copyright Office circulars.
work_coding_tag_written_instruct work_coding_tag_written_recipe work_coding_tag_written_rules work_coding_tag_written_referenc work_coding_tag_written_menu work_coding_tag_written_phrase work_coding_tag_visual_photo_pro work_coding_tag_visual_logo work_coding_tag_visual_icon work_coding_tag_software_webdesi work_coding_tag_software_scraped work_coding_tag_extra_idea work_coding_tag_extra_pitch work_coding_tag_extra_secret work_coding_tag_extra_patented work_coding_tag_extra_trademark work_coding_tag_extra_key work_coding_tag_extra_jewelry work_coding_tag_extra_clothing link_coding_claim_trademark link_coding_claim_secret link_coding_claim_patent link_coding_claim_private link_coding_claim_defamation link_coding_claim_anti_circumven link_coding_claim_scraping link_coding_claim_counterfeit link_coding_claim_other link_coding_inferred_trademark link_coding_inferred_secret link_coding_inferred_patent link_coding_inferred_private link_coding_inferred_defamation link_coding_inferred_anti_circum link_coding_inferred_scraping link_coding_inferred_counterfeit link_coding_inferred_other link_coding_tag_extra_key

Other_que **Other** As noted above, this composite variable is a "catch-all" for other variables that came up relatively infrequently. It includes requests targeting material that is embedded on a page (and thus hosted elsewhere); requests targeting material where the site where the material is posted or the alleged direct infringer appear to be competitors of the principal; requests where the alleged infringer appears to be the owner of the AIW; requests where the alleged infringer appears to be authorized in posting or using the AIW; and requests that indicate questions related to the ownership of the AIW.
link_coding_market_hot_link analysis_target_competitor analysis_infringer_competitor analysis_infringer_owner analysis_infringer_authorized work_coding_copyright_holder_sub work_coding_copyright_holder_inc notice_coding_sender_role_unconn
Que_total **Total Questionable** This composite variable includes all "questionable" requests.
SR_que FU_que SM_que Other_que
Total_PP **Total Potentially Problematic** This composite variable includes all potentially problematic requests, including both the "questionable" requests and the fundamentally flawed "No Match" requests.
SR_que FU_que SM_que Other_que link_coding_amount_none

C. STUDY 3

1. Sampling

Study 3's findings are based on a random sample of 1732 requests, drawn from the subset of 33,409 requests[376] in the six-month dataset that were sent to Google Image Search (i.e. requests where the "notice_re_line" is "Infringement Notification via Imagesearch

[376] This number is an approximation because of the manner in which structured data from Google's Image Search complaint web form interacted with our coding engine algorithm. For Image Search, Google requests three separate URLs for one instance of infringement: a link to search results page, a hotlink to hosted image file, and a link to the page where the file appears. The data structure did not allow us to distinguish between this situation and a notice that contains multiple separate takedown requests URLs and unfortunately, senders often did not comply with the three-link request, leaving us without a simple calculation to estimate unique requests in the entire Google Image Search set. We did, however, record when a sender in our sample provided more than one URL to identify the alleged infringement. Based on our coded sample, only 15.2% of the non-Miller requests provided more than one URL to identify the alleged infringement (with 7% providing all three URLs requested by Google). We note that nearly two-thirds of these are from just two senders. Finally, the full count of 33,409 requests includes a number of requests that our coding engine algorithm improperly identified as AIMs. These incorrectly selected requests were excluded from our sample and analysis.

Accordingly, the margin of error for this sample—+/-2.29 at a 95% confidence interval—is artificially inflated because it is based on the full count of 33,409 requests. Because some links are irrelevant repeats of the same request in different form, the sample is actually a larger proportion of the full count than it first appears, making this is a conservatively wide margin of error.

Complaint"). This gives us an approximate margin of error of +/-2.29 at a 95% confidence interval.[377]

As with the Study 2 data, a randomization function presented these takedown requests to reviewers in randomized order.

2. Additional "Clean-Up" Processing

The idiosyncratic nature of the Google Image Search takedown requests required more attention in the cleaning phase than the relatively standard Google Web Search requests that predominated in the Study 2 sample. We systematically reexamined several descriptive variables before reporting:

- Industry of Principal: The categories established for the Google Web Search sample were not always sufficient for the Google Image Search set. In particular, we noted that a large number of the Google Image Search sender's industries were originally categorized as "other". To address this, we created new variables for "Art/Design" ("notice_coding_prin_indu_art") and "E-Commerce" ("notice_coding_prin_indu_ecomm"), and then systematically recategorized requests to ensure consistency across the sample.

- **Type of Target Site (Miller Requests):** As noted on pages 99-100 of the report, one sender, the pseudonymized "Ella Miller," sent more than half of all the Google Image Search requests. A large number of Miller requests marked as "analysis_target_role_other" were targeting message boards, a target role for which, in an oversight, we had not created a specific variable. To allow for quantification of this prominent feature, relevant requests were removed from "other" and recategorized as "analysis_target_role_fan_site" (the "analysis_target_role_fan_site" variable was created for fan sites or forums). Moreover, if Miller requests were marked as "analysis_target_role_blog", then these requests were removed from "analysis_target_role_social" to avoid duplication.
  - Relatedly, we note our original oversight not to create separate variables for "fan site" and for "forum". Instead, these types of target sites were captured together in the "analysis_target_role_fan_site" variable, which was used to indicate fan sites or forums (used by fans or not). Accordingly, please note: researchers who would like to distinguish between these two types of target sites will need to recode these requests to account for the difference.

- **Substantive Questions:** We found we could not always use the same individual variables that we combined in Study 2 to evaluate substantive issues for this set. This is because we found that the vagaries of the Study 3 data meant that these variables—which would otherwise be relevant to copyright claims—sometimes generated false positives for the particular substantive issue we were examining.

  For example, 1122 requests in the Google Image Search set were marked as user-generated content ("link_coding_tag_extra_user"). Normally, this would be a rough indication that the challenged material was relatively likely to be transformative and relatively unlikely to affect the market for the original work, making it reasonable to consider fair use. Because the senders of many of these particular 1122 takedown requests appeared to be making non-copyright claims, however, the fact that user-generated content may be fair use under copyright law was not relevant to the claim.

[377] Due to the characteristics of the Google Image Search dataset, this margin of error is artificially inflated. The actual margin of error is likely narrower. Please see footnote 7 for the complete explanation of this issue.

We realize this is a judgment call—it could be considered defensible to count all of the potential flaws in a request using the assumption that it was essentially claiming copyright infringement, regardless of whether the underlying claim was actually close to copyright or not. Nonetheless, we chose a more conservative approach and considered such a result to be a false positive. We therefore built a separate set of variables to examine substantive issues that arose in this set.

Note: As a side effect of this process, composite variables do *not* include otherwise potentially relevant individual variables that appeared in zero notices. This is different from Study 2, where composite variables simply included all potentially relevant variables, even if the count for any individual variable was zero. Researchers should be aware of this difference, and that it is a side effect of managing the (in our view) false positive results that would otherwise come up in Study 3.

3. Errata

There were a few minor errors in the text of the original published report:

- Page 101: Total "Individual" principals should be 68.9% (reported as 68.8%).
- Figure 12 (page 103): Sender's Country "Other" should be 9% (reported as 9.1%).
- Page 104 (and Figure 13, page 105): Total type of target sites that are "News" should be 7.4% (reported as 7%).
- Page 108: Requests which raised trademark concerns should be 2.1% (reported as 1.8%).
- Page 108: Requests which presented a fair use question should be 11.5% (reported as 11.6%).
- Page 110 (and Figure 15): Targeting material in the public domain should be .9% (reported as 1%).
- Page 146: (and Figure 18, page 147): Visual AIW should be 97.9% (reported as 98%).
- Questionable (including Miller): 70% (reported as 70.2%).
- Questionable (excluding Miller): 36.3% (reported as 36.8%).

4. Study 3: Variable Counts

CATEGORY	VARIABLE(S) USED	COUNT	PERCENT
Figure 8: Sender Role (Google Image Search Sample) (page 100)			
Principal	notice_coding_sender_role_princi	Total: 1635 Non-Miller: 719 Miller: 916	Total: 94.4% Non-Miller: 88.1% Miller: 100%
Unknown	notice_coding_sender_role_unknow	Total: 88 Non-Miller: 88 Miller: 0	Total: 5.1% Non-Miller: 10.8% Miller: 0%
Agent	notice_coding_sender_role_agent	Total: 9 Non-Miller: 9 Miller: 0	Total: .5% Non-Miller: 1.1% Miller: 0%
Figure 9: Sender Size (Google Image Search Sample) (page 101)			
Individual	notice_coding_principal_individu	Total: 1193 Non-Miller: 277 Miller: 916	Total: 68.9% Non-Miller: 33.9% Miller: 100%
Small Business	notice_coding_prin_bus_small	Total: 335 Non-Miller: 335 Miller: 0	Total: 19.3% Non-Miller: 41.1% Miller: 0%
Small-Medium Business	notice_coding_prin_bus_sm_med	Total: 6 Non-Miller: 6 Miller: 0	Total: .3% Non-Miller: .7% Miller: 0%

Medium Business	notice_coding_prin_bus_med	Total: 2 Non-Miller: 2 Miller: 0	Total: .1% Non-Miller: .2% Miller: 0%
Unknown	N/A - Remainder	Total:196 Non-Miller: 196 Miller: 0	Total:11.3% Non-Miller: 24% Miller: 0%
Figure 10: Industry of Principal (Google Image Search Sample; Non-Miller Requests) (page 102)			
E-Commerce	notice_coding_prin_indu_ecomm	42	5.1%
Other/Unknown	N/A	51	6.3%
Photography	notice_coding_prin_indu_photo	86	10.5%
Individual	notice_coding_principal_individu	277	33.9%
Adult Entertainment	notice_coding_principal_ind_adul	166	20.3%
Art/Design	notice_coding_prin_indu_art	194	23.8%
Figure 11: Sender's Country (Google Image Search Sample; Non-Miller Requests) (page 103)			
United States	notice_sender_country is US	355	43.5%
Other	notice_sender_country is anything except US, DE, IN, IL, GB, or CN	157	19.2%
Germany	notice_sender_country is DE	134	16.4%
India	notice_sender_country is IN	61	7.5%
Israel	notice_sender_country is IL	61	7.5%
Great Britain	notice_sender_country is GB	27	3.3%
China	notice_sender_country is CN	21	2.6%
Figure 12: Sender's Country (Google Image Search Sample; Including Miller Requests) (page 103)			
Sweden	notice_sender_country is SE	917	52.9%
United States	notice_sender_country is US	355	20.5%
Other	notice_sender_country is anything except SE, US, DE, IN, IL, GB, or CN	156	9%
Germany	notice_sender_country is DE	134	7.7%
India	notice_sender_country is IN	61	3.5%
Israel	notice_sender_country is IL	61	3.5%
Great Britain	notice_sender_country is GB	27	1.6%
China	notice_sender_country is CN	21	1.2%
Table 3: Percentage of Requests from High-Volume Senders in Google Image Search Request (page 103)			
Ella Miller	notice_sender_attorney is Ella Miller	916	52.9%
Lil Duck Duck	notice_sender_attorney is Lisa Eller	148	8.5%
Purzel-Video GmbH	notice_sender_attorney is Martin GÌ?bel (89) or Rosanna Schilling (17)	106	6.1%
Anatoli Ivanov	notice_sender_attorney is Anatoli Ivanov	61	3.5%
Alpha Sky Productions	notice_sender_attorney is Steve Westin	54	3.1%
Ragalahari	notice_sender_attorney is Suresh Maganti	51	2.9%
Oasis Costumes	notice_sender_attorney is Xiongbing Wan	20	1.2%
Figure 13: Types of Target Sites (Google Image Search Sample; Non-Miller Requests) (page 105)			
Other or Unknown	analysis_target_role_other	389	47.7%
Social Media	analysis_target_role_social	201	24.6%
Personal Website/Blog	analysis_target_role_blog	127	15.6%
Individual	analysis_target_role_individual	89	10.9%
News	analysis_target_role_news	60	7.4%
E-Commerce	analysis_target_role_ecommerce	56	6.9%
Fan Site/Forum	analysis_target_role_fan_site	51	6.3%
Video Streaming	analysis_target_role_video	50	6.1%
Torrent	analysis_target_role_torrent	48	5.9%

Aggregator	analysis_target_role_aggregator	31	3.8%
Cyberlocker	analysis_target_role_cyberlocker	14	1.7%
Corporate	analysis_target_role_corporate	13	1.6%
File Search	analysis_target_role_file_search	8	1.0%
Education	analysis_target_role_educational	7	.9%
Figure 14: Types of Target Sites (Google Image Search Sample; Miller Requests) (page 106)			
Fan Site/Forum	analysis_target_role_fan_site	720	78.6%
Personal Website/Blog	analysis_target_role_blog	140	15.3%
Individual	analysis_target_role_individual	133	14.5%
Cyberlocker	analysis_target_role_cyberlocker	22	2.4%
News	analysis_target_role_news	16	1.7%
Social Media	analysis_target_role_social	8	.9%
Other or Unknown	analysis_target_role_other	7	.8%
Section IV.C.2.b.i **Nearly One in Six Requests Raised Questions about the Subject Matter of the Claim (page 108)** Please see Section B.5 above for further description of our methodology for identifying characteristics could raise subject matter questions. We used the same basic methodology here; however, to avoid false positives, we modified our composite variables as described above in Section C.2. Accordingly, only those characteristics which actually showed up in the dataset are included in this composite. Again following our general methodology, requests that exhibited more than one of these variables were only counted once.			
Subject matter other than copyright (total)	work_coding_tag_extra_trademark; work_coding_tag_visual_logo; work_coding_tag_visual_photo_pro; link_coding_claim_counterfeit; link_coding_claim_defamation; link_coding_claim_other; link_coding_claim_private; link_coding_claim_trademark; link_coding_inferred_counterfeit; link_coding_inferred_defamation; link_coding_inferred_other; link_coding_inferred_private; link_coding_inferred_trademark	123	15.1%
Privacy	link_coding_claim_private; link_coding_inferred_private	49	6%
Product Photos	work_coding_tag_visual_photo_pro	49	6%
Trademark	work_coding_tag_extra_trademark; link_coding_claim_trademark; link_coding_inferred_trademark	17	2.1%
Defamation	link_coding_claim_defamation; link_coding_inferred_defamation	12	1.5%
Other (e.g. online impersonation; harassment)	link_coding_claim_other; link_coding_inferred_other	11	1.3%

Section IV.C.2.b.ii One in Nine Requests Presented a Fair Use Question (page 108) Please see Section B.5 above for further description of our methodology for identifying characteristics that would potentially weigh in favor of fair use. We used the same basic methodology here; however, to avoid false positives, we modified our composite variables as described above in Section C.2. Accordingly, only those characteristics which actually showed up in the dataset are included in this composite. Again following our general methodology, requests that exhibited more than one of these variables were only counted once.			
Characteristics that weigh favorably toward fair use	work_coding_tag_extra_educationa; link_coding_amount_lt50; link_coding_amount_small; link_coding_market_comments; link_coding_market_context; link_coding_market_criticize; link_coding_market_curated; link_coding_market_other_comment; link_coding_tag_extra_educationa; link_coding_tag_extra_news; link_coding_tag_visual_photo_new; link_coding_tag_visual_photo_thu; analysis_target_educational; analysis_target_role_educational; analysis_target_role_news	94	11.5%
Section IV.C.2.b.iii Ownership Issues (page 109)			
Sender appears to be the subject of the photograph	work_coding_copyright_holder_sub	50	6.1%
Sender appears to be the subject of the photograph (excluding apparent self-portraits)	4 instances where it appeared that the photograph was a self-portrait were set aside for this count	46	5.6%
Section IV.C.2.b.iv Only a Small Percentage of Image Search Requests Failed to Identify Clearly the Works in Question (page 109) Following our general methodology, requests that exhibited more than one of these variables were only counted once.			
Problem identifying the work in question	link_coding_unavailable_aggregat; link_coding_unavailable_search_r; link_coding_unavailable_multiple; link_coding_unavailable_navigati; work_coding_id_no	24	2.9%
Section IV.C.2.b.iii Public Domain (page 110) **Please note:** to avoid false positives, we modified our composite variables as described above in Section C.2. Accordingly, only those characteristics which actually showed up in the dataset are included in this composite. This is why the variable for United States government works is not listed (U.S. government works are in the public domain, but no U.S. government works showed up in response to our query). We further note that state government works vary in terms of whether the state claims copyright. We did not investigate deeply enough to determine whether state works emanated from states that claim copyright—as with other categories, our goal was to find out whether there were takedown requests that raised questions that would militate toward further review. Please also note that we did not include foreign government works, as foreign governments commonly claim copyright. Requests that exhibited more than one of these variables were only counted once. Finally, please note that, at about 1%, this category is within the margin of error and can only be considered anecdotal.			
Targeting material that is potentially in the public domain	work_coding_tag_extra_pre_1923; work_coding_tag_extra_state	7	.9%
Total Questionable Requests with more than one of these variables were only counted once.			
Total Questionable (with Miller requests)	N/A—All Miller requests plus 296 non-Miller requests (as detailed below)	1212	70%
Total Questionable (without Miller requests)	All variables indicated above plus work_coding_copyright_holder_inc	296	36.3%
Appendix A; Figure 18: **Type of Media Alleged to Be Infringed in Takedown Requests (Google Image Search Sample; Non-Miller Requests) (page 147)** These numbers add up to more than 100% because some content fell into more than one category.			
Visual	work_coding_tag_visual	799	97.9%
Written	work_coding_tag_written_work	138	16.9%
Audiovisual	work_coding_tag_av	2	.2%
Software	work_coding_tag_software	1	.1%

Appendix A; Figure 19: Type of Allegedly Infringing Material (Google Image Search Sample; Non-Miller Requests) (page 147) These numbers add up to more than 100% because some content fell into more than one category.			
Visual	link_coding_tag_visual	704	86.3%
Written	link_coding_tag_written_work	113	13.8
Audiovisual	link_coding_tag_av	27	3.3%
Software	link_coding_tag_software	1	.1%

Made in the USA
Monee, IL
08 November 2026

67984630R00103